Acknowledgements

We wish to thank our Secretariat for their assistance in our work. Our Secretary, Dr Heather Walton, has made an outstanding contribution: her ability to synthesise epidemiological and economic perspectives has been particularly helpful. We also thank Mr John Henderson for his expert contribution to the economic parts of our report. Thanks also to Miss Julia Cumberlidge for general secretarial help. Additional thanks go to: Professor Paul Jones (St. George's Hospital Medical School) for his help regarding the St George's Respiratory Questionnaire; Professor Ross Anderson (St. George's Hospital Medical School) for early discussion of the epidemiological evidence for the sensitivity analysis; and Dr Peter Bennett (Department of Health) for his discussion papers on various policy appraisal approaches. The guidance we received from Professor Peter Burney and Mr Fintan Hurley is acknowledged in Appendix 2.

Permission to reproduce the EUROQOL EQ5D Questionnaire and Table 3A.1 in Annex 3A has been applied for.

DEPARTMENT OF HEALTH

Ad-Hoc Group on the Economic Appraisal
of the Health Effects of Air Pollution

ECONOMIC APPRAISAL OF THE HEALTH EFFECTS OF AIR POLLUTION

London: The Stationery Office

Published with the permission of the Department of Health on behalf of the Controller of Her Majesty's Stationery Office.

© Crown Copyright 1999

All rights reserved.

Copyright in the typographical arrangement and design is vested in the Crown. Applications for reproduction should be made in writing to the Copyright Unit, Her Majesty's Stationery Office, St Clements House, 2–16 Colegate, Norwich NR3 1BQ.

First published 1999

ISBN 0 11 322272 6

Contents

page

Executive Summary 1

Chapter 1: Introduction 9

Chapter 2: Quantification of Health Effects 13

Chapter 3: Air Pollution Policy Appraisal 29

Chapter 4: Benefits of Lower Mortality Risks 59

Chapter 5: Benefits of Reduced NHS Costs and Other Costs 83

Chapter 6: Benefits of Less Morbidity 89

Chapter 7: Summary and Conclusions 103

Appendix 1: Glossary of Terms and Abbreviations 123

Appendix 2: Membership of the Ad-Hoc Group on the
Economic Appraisal of the Health Effects of Air Pollution 127

Appendix 3: Members' Interests 129

Executive Summary

1. The United Kingdom (UK) National Air Quality Strategy sets out objectives for further reductions in the levels of air pollutants by the year 2005. This report considers how assessment of the health benefits of these reductions might be undertaken and covers all the pollutants in the strategy except benzene, 1,3-butadiene and lead. The conclusions will feed into the consideration of overall costs and benefits in the first Strategy review due to be completed by the end of 1998.

Terms of reference

2. This report has been prepared by an Ad-Hoc Group on the Economic Appraisal of the Health Effects of Air Pollution set up by the Department of Health with the following terms of reference:

   a) to advise on how best to reflect the importance of health effects in any cost/benefit decisions in air quality policy;

   b) to estimate the health care cost implications of changes in levels of air pollution;

   c) to consider whether monetary valuation of health effects is appropriate in this context;

   d) to consider the merits of alternative approaches and, if necessary, to recommend further work to help develop the most appropriate approach;

   e) if possible, to produce estimates of costs (with gaps in information and assumptions clearly stated) and, if necessary, to recommend further work to improve these estimates.

3. This report primarily addresses methodological issues and is very much a first step in an area which has not previously been considered in any detail.

Health effects

4. In January 1998, the Committee on the Medical Effects of Air Pollutants (COMEAP) published a report on the quantification of the health effects of air pollutants in the UK. The evidence for the effect of particles, sulphur dioxide and ozone on deaths brought forward and on respiratory hospital admissions was considered sufficiently robust to be used for quantification. It is important to recognise that the deaths are thought to occur mainly in the elderly with advanced lung or heart disease and to be brought forward by weeks or months but not years, although the loss of life expectancy is not known precisely. Air pollution acts as an aggravating rather than a causal factor and this has implications for judging the importance of its effects.

5. Air pollution also has other health effects for which the evidence is less strong. We discuss whether some of this evidence could be examined in sensitivity analyses to give a rough idea of the *possible* impact on the overall results. We emphasise that, if this is done, the major uncertainties in the interpretation of these results should always be stated.

Assessing the importance of the health benefits - our approach

6. Some measure is needed of the importance of the quantified health effects so that they can be compared with the other implications of the policy concerned. Individuals', experts' and society's views on this may differ. There is no simple rule on which perspective *should* be used but we note where the perspectives differ in our discussion of possible approaches.

7 Society needs to ensure that large amounts of money are not spent on trivial risks when the money could be better spent on more significant risks i.e., the costs should not exceed the benefits. To compare the costs and benefits directly, the benefits need to be in the same units as the costs. A monetary value for the benefits that reflects the preferences of those at risk can be obtained by finding out what they would be willing to pay to reduce a particular risk. Although reductions in risks are typically not marketable goods, people do pay for measures to reduce risks either directly or through taxation and people do trade off small risks against other things which are important to them. These trade-offs can be investigated in carefully designed surveys.

8 These monetary values may be expressed as the "value of preventing a statistical fatality". Suppose people are, on average, willing to pay £10 for a safety improvement which will reduce their individual risk of death during the coming year by 1 in 100,000. This risk reduction would mean, on average, that, in a group of 100,000 people, there would be 1 fewer premature deaths. These 100,000 people would, between them, be willing to pay £10 x 100,000 = £1 million where, on average, or "statistically", 1 fatality would be prevented. This is not about valuing particular individuals' lives but about what people, collectively, are willing to pay for a small reduction in the risk faced by each of a large number of individuals.

9 There have been some criticisms of the techniques for determining individual willingness to pay (WTP) for risk reduction, but we consider that the well-conducted studies do give an adequate indication of the broad order of magnitude of the values people place on reducing risks.

10 The benefits of risk reduction could also be expressed in terms of standard measures of health gain such as improvements in quality of life and life expectancy. This would assist comparison with other public health interventions. One possible standard measure is to use quality-adjusted life years (QALYs) where quality of life (measured on a scale from 0 (death) to 1 (full health)) is used to weight the appropriate number of years (e.g., 2 years with a quality of life of 0.5 would represent 1 quality-adjusted life year). The costs per quality-adjusted life year gained can then be determined. This measures cost-effectiveness but does not show whether the benefits exceed the costs. WTP will differ at different ages and for different diseases but quality-adjusted life years, by convention, do not.

11 Most policy decisions involve weighing several factors in an informal way. The same thing can be done more formally in multi-criteria analysis, which can help structure the process. The method lacks, however, the advantage of a single encompassing rule, which converts all impacts into monetary values and estimates the net benefits of the policy options as could be done with cost-benefit analysis. Cost-benefit analysis information can be included within a multi-criteria analysis so in this case it would be a complementary rather than alternative approach. It can also be used with the public to obtain more information on people's qualitative views on the importance of the benefits.

12 We concluded that, in the longer term, the most fruitful approach may be to use a combination of approaches and cross-check the results obtained. We considered that monetary valuation most clearly demonstrates whether the benefits exceed the costs but that expressing benefits in terms of quality of life and life expectancy would be helpful for comparison with other health interventions. There may be debate about the use of these approaches but we emphasise that they do not provide the only evidence on which a decision will be based.

Benefits of lower mortality risks

Gains in quality of life and life expectancy

13 The average loss of life expectancy when deaths are brought forward by air pollution is unknown but, from clinical judgement, is unlikely to be more than between a month and a year for respiratory deaths. Limited evidence suggests that the quality of life score could range from 0.2 to 0.7 for patients with advanced chronic obstructive pulmonary disease (COPD). Thus, if air pollution did not contribute to an earlier death in one of these patients, there might be a gain of up to about 0.02-0.7 quality-adjusted life years.

Monetary valuation

14 There are no suitable studies of WTP for reductions in air pollution mortality risks. There are WTP studies for reductions in risks of deaths in road accidents. However, we do not consider these results can be applied to the air pollution context without adjustment because we believe that the nature of the risks and the characteristics of those affected are too different. From an understanding of the factors which affect WTP we can suggest what values might be expected in the air pollution context were those at risk to be asked (this includes those who are not currently in the group at risk but may be at risk in the future). This approach is of course speculative and only an interim solution.

15 People generally perceive air pollution risks as involuntary, poorly understood and uncontrollable but perceive road accident risks as largely voluntary, well-understood and easy to control. Empirical evidence suggests that WTP for reductions in such involuntary risks can exceed that of voluntary risks by a factor of up to 2 or 3. The WTP component of the value of preventing a statistical fatality (VPF) for road accident deaths used by the Department of the Environment, Transport and the Regions (DETR) is £0.8 million so, for an involuntary risk from air pollution, the VPF might approach £2 million. Alternatively, the average VPF for people of average age (about 40) and average health in various contexts is about £2 million. We therefore took £2 million as the baseline from which to make adjustments to allow for the age and impaired health of those affected by air pollution. We recommend the following adjustment factors:

(i) those affected by air pollution are mainly over 65 and WTP in this age group as a whole is about 70% of that at the average age of 40;

(ii) those affected have a lower life expectancy than average for their age. We would expect this to further reduce WTP but it is unclear by how much. Theory and understanding of the relationship of WTP with age does not suggest that WTP would drop more than in proportion to life expectancy. Thus, for a life expectancy of 1 year rather than 12 years (the average for over 65s), WTP would be reduced to one twelfth and for a life expectancy of 1 month by up to a further factor of twelve;

(iii) those affected also have a lower quality of life (0.2 to 0.7) than average for their age (0.76). Again, we would expect this to reduce WTP since people seem likely to value further time in full health more highly than further time in poor health. Given the lack of good evidence, we suggest WTP might reduce in proportion to quality of life i.e., to reduce by 0.2/0.76 or 0.7/0.76 for a quality of life of 0.2 or 0.7, respectively;

(iv) a small reduction in risk may be more important to people who already have a high risk of death but we consider the baseline level of risk in the general elderly population is insufficiently high to warrant adjustment. However, there is some debate over whether the baseline risk in those with advanced disease (unknown) might be high enough to result in an increase in WTP and, if so, how significant this increase might be. On equity grounds we do not suggest adjustment for income.

16 The result of taking these adjustments together and applying them to the £2million baseline is shown in Figure 1.

We acknowledge there are ethical arguments for and against the adjustments. Although WTP clearly does decline in the elderly, there is more uncertainty over the other adjustments and these have a large potential impact. We also note the debate over whether the possible effect of the high baseline level of risk on WTP might lessen the impact of these adjustments. Due to the uncertainty, we suggest a wide range of estimates. We would expect the WTP to reduce air pollution mortality risks to be below £1.4m (no adjustment for impaired health) and possibly to be as low as £2,600 for a loss of life expectancy of 1 month and a quality of life of 0.2. We note that the lower end of the range is substantially lower than the VPF for road accidents but believe this is reasonable given the smaller losses in quality of life and life expectancy involved.

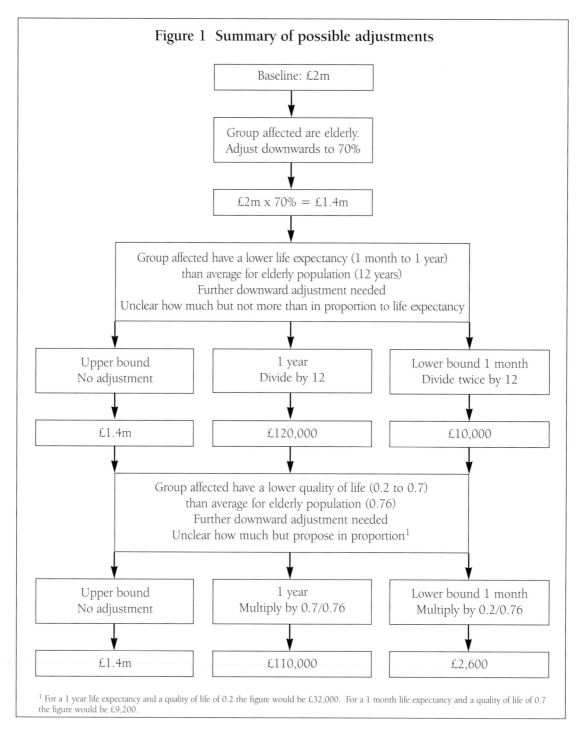

Figure 1 Summary of possible adjustments

[1] For a 1 year life expectancy and a quality of life of 0.2 the figure would be £32,000. For a 1 month life expectancy and a quality of life of 0.7 the figure would be £9,200.

NHS costs

17 If people's lives are extended by perhaps 1 month to 1 year by reductions in air pollution, there could be additional costs of about £200 to £2500 to the NHS. In financial terms, we believe it is reasonable to present NHS savings net of these costs but acknowledge that the extent to which this is taken into account in policy-making is controversial. We present results with and without these costs.

Chronic exposure and mortality

18 There is evidence from the United States (US) for a chronic effect of particles on mortality although it is uncertain whether this evidence would apply in the UK and, if so, what the size of the impact would be. The chronic effects were not quantified by COMEAP for these reasons

although it was noted that the impact was potentially substantial. We have not had time to explore the valuation aspects fully (we did not discuss modifications of the value of life years lost approach, for example) but consider that, in the interim, our adjustment factor approach may reasonably be applied provided sufficient information is available on the characteristics of those affected. If those affected are not already in poor health, the WTP values could, potentially, be higher than for acute deaths.

Benefits of less morbidity

19 The benefits of reduced morbidity are made up of reductions in:

   (i)   NHS costs;

   (ii) private costs e.g., travel to the doctors (small); costs of avertive behaviour (unknown);

   (iii) lost output if people are prevented from working through ill-health (unlikely in the elderly who already have advanced disease for other reasons);

   (iv) welfare costs (reflecting the pain and discomfort of illness).

NHS costs

20 The average cost for all ages for a respiratory hospital admission is £1,400. Air pollution has a greater effect on the elderly so the cost could be nearer to that for admissions in the over 65s (£2,500). The average durations are 8 and 14 days, respectively. Other NHS costs are likely but there is limited information on these.

Welfare costs

Gains in quality of life

21 Reductions in respiratory hospital admissions associated with air pollution would not return patients with underlying disease to full health. We need a measure of the *shift* in quality of life influenced by air pollution. There are no good data on this. Limited evidence suggests a possible shift of perhaps between 0 and 0.8. For a duration of hospital stay of about 8-14 days this corresponds to 0-0.03 quality-adjusted life years per admission.

Monetary valuation

22 There are no WTP studies available for avoiding a respiratory hospital admission although there are for some respiratory symptoms. We considered whether a rough estimate could be inferred from other studies. One option is the rule of thumb derived from US data that WTP is roughly twice the cost of illness. This has little plausible basis (some serious diseases cost very little to treat as they are untreatable).

23 Another option is to use an equation correlating WTP and quality of well being (QWB) scores. This equation suggests that the same shift in quality of life is valued differently for different baseline QWB scores. QWB scores of 0.5-0.7 have been obtained from COPD patients and we suggest a QWB score of 0.47 during a hospital admission. The equation predicts a WTP of £170-£735 to avoid a shift in QWB score from 0.5-0.7 to 0.47 for a duration of 8-14 days. This is very uncertain. The equation is based on theoretical quality of life scores for minor respiratory symptoms and may not apply to more serious effects. The answer is very dependent on the quality of life data which is itself limited. This range is lower than other figures used in the literature. Many of these are based on the cost of illness rule of thumb which is not firmly based. None are based on direct evidence. It is possible, given the uncertainties, that the value could be higher than we have predicted but this can only be resolved by new empirical evidence.

Conclusions

24 The health benefits probably dominate the overall benefits of reductions in air pollution so it is important to try to include them in a cost-benefit analysis if possible. Our suggested estimates are summarised in Table 1 below along with some associated uncertainties. Our predicted estimates for mortality risks are lower than used by others but we believe this is appropriate given that air pollution is thought to act as an aggravating factor bringing deaths forward rather than being the entire cause of the deaths. Nonetheless, there are substantial uncertainties in the exact size of the estimates. We emphasise these are very much interim estimates and direct evidence is needed (see below).

Table 1 **Summary of Estimates of Measures of Importance of Benefits from Reductions in Air Pollution (1996 Prices)**

Measure of importance of benefit from reduction in pollution	Per death brought forward (acute)	Per respiratory hospital admission	Comments
Gains in quality of life and life expectancy	Quality of life of 0.2 to 0.7 for perhaps 1 month to 1 year. (0.02 - 0.7 QALYs) (respiratory)	Avoided reduction in quality of life of 0 - 0.8 for 8 - 14 days (0 - 0.03 QALYs)	For hospital admissions, *change* in quality of life due to air pollution not directly studied - inferred from some empirical evidence on COPD patients before hospital admission.
NHS costs saved	None saved, added costs in extra 1 month to 1 year of life: £200 - £2500.	£1,400 - £2,500	In-patient costs only. There could be other NHS costs.
People's willingness to pay (WTP) for a small reduction in risk aggregated so as to apply per death brought forward or for avoiding a hospital admission	Upper bound: £1.4m Low estimates: 1 year £32,000 to £110,000 1 month £2,600 to £9,200 (respiratory)	£170 - £735	No direct empirical evidence. Mortality value predicted from judgements of effect of context, age and impaired health state on WTP. Respiratory hospital admission figure from relationship between WTP and quality of life using empirical evidence and predicted scores. Other inferred figures in the literature are higher.
Other		[Private costs unknown but small]	Savings in costs of avertive behaviour unknown. Lost work output n/a to elderly or seriously ill.

25 These interim estimates can be used in conjunction with the dose-response functions from the COMEAP report (paragraph 4) to give an estimate of benefits per unit concentration of pollutant (see Table 2).

Table 2 **Summary of Estimates of Measures of Importance of Benefits per Unit Concentration of Pollutant per Year or per Summer**

	Health outcome[a] and measure of importance[b] per $\mu g/m^3$ reduction of pollutant		
	PM_{10} in GB urban population per year	Sulphur dioxide in GB urban population per year	Ozone in GB urban and rural population summer only
Reduction in deaths brought forward (all cause) (acute)	340 (270 - 385)	270 (225-315)	170 (70-225)
WTP for reduction[c]	£0.7m - £540m	£0.58m - £440m	£0.18m - £315m
Reduction in respiratory hospital admissions	280 (180-390)	180 (-53 to 320)	145 (50 - 245)
NHS savings	£0.25m - £0.98m	-£0.07m - £0.8m	£0.13m - £0.61m
WTP for reduction	£0.03m - £0.29m	-£0.01m to £0.24m	£0.01m - £0.18m
Total NHS savings	£0.25m - £0.98m	-£0.07m - £0.8m	£0.13m - £0.61m
Total WTP	£0.73m - £540m	£0.57m - £440m	£0.19m - £315m
Total benefits	**£0.98m - £540m**	**£0.5m - £440m**	**£0.32m - £315m**
NHS costs incurred[d]	£0.05m - £0.96m	£0.05m - £0.79m	£0.01m - £0.56m
Total benefits net of NHS costs	**£0.93m - £540m**	**£0.45m to £440m**	**£0.31m - £315m**

Notes to Table 2:

[a] Calculated from the mean and standard deviation of the dose-response functions (percentage increase in health outcomes per unit concentration of pollutant) in the COMEAP report, baseline rates for the health outcomes and numbers in the relevant population (see Chapter 2 and Annex 2A). Figures would be higher if the Northern Ireland population was included and lower for hospital admissions associated with ozone if there is a threshold of 50 ppb.

[b] Given as range from highest estimate (top end of range for health outcome and top end of relevant range in Table 1) to the lowest estimate (bottom end of the relevant ranges). The figure of -53 for hospital admissions and sulphur dioxide represents statistical variability - a beneficial effect of sulphur dioxide is not supported by other evidence.

[c] Upper figure based on upper bound VPF of £1.4 m; probably an overestimate. Lower figure based on proportional adjustment according to a loss of life-expectancy of 1 month and a quality of life of 0.2.

[d] NHS costs incurred if lives extended by 1 month to 1 year.

This indicates that the top end of the range for WTP for reductions in risks of deaths brought forward dominates the overall results. It also shows that reductions in levels of particles could result in the greatest benefits per unit concentration of pollutant. However, we emphasise that this does not take into account the actual concentrations to which people are exposed. It should also be noted that the WTP estimates do not represent actual costs incurred.

26 The Group was concerned that the quantified effects might represent only a fraction of the overall effects of air pollution. In particular, the chronic effects of particles, if confirmed, could dominate the results. Also, WTP to reduce risks of cardiovascular deaths (which are included in the deaths brought forward) might be higher as patients could be younger with a higher quality of life.

27 Morbidity effects have a less significant impact on overall benefits but are also only partially covered. Air pollution may influence cardiovascular admissions - the NHS costs and WTP values could be less or more than for respiratory admissions depending on the type of heart disease affected. This is unclear. The possible effect of air pollution on more minor symptoms could have a significant impact if the numbers of people affected were very much larger than for the more serious effects. This is unknown.

28 In addition, not all pollutants have been covered. Nitrogen dioxide and carbon monoxide were considered in the COMEAP report but the evidence for effects was not strong enough for full quantification. Benzene, 1,3-butadiene and lead were not included in the COMEAP report and there are some difficulties in quantifying their effects. As a general point, we note that the absence of quantified benefits should not necessarily be taken to mean there are no health benefits. We hope that further work will clarify the extent of these additional non-quantified benefits.

Further work

29 We support the further epidemiological work recommended in the COMEAP report. We also recommend work on the following:

people's attitudes to air pollution risks;

quality of life in patients with exacerbations of cardiovascular and respiratory disease;

WTP to reduce the risk of deaths brought forward by air pollution;

WTP for reductions in respiratory hospital admissions and other morbidity effects;

the valuation of chronic effects;

investigation of private costs and lost output;

the different factors influencing WTP;

the relationship between WTP and quality of life.

Chapter 1
Introduction

Air pollution - why is it important?

1.1 Throughout history, some human activities have polluted the air. These activities have advantages such as the warmth from burning fuel, the increase in people's quality of life from the products of industrial activity and the freedom of car travel. However, the pollution generated can have detrimental effects on people's health and on the environment and the people affected may not be the same people as those enjoying the advantages of the original activity[1]. This report addresses the importance which should be placed on the health effects of air pollution when balancing the advantages and disadvantages of polluting activities.

1.2 The effects of air pollution on health have been most clearly seen during acute and severe air pollution episodes. For example, in London in December 1952, very high concentrations of smoke and sulphur dioxide over several days led to over 4000 excess deaths (Ministry of Health, 1954). Domestic coal burning was a major contributor to air pollution at that time. Levels of air pollution have since declined significantly and traffic is now the major source of air pollution in cities. Although the health effects of air pollution are not as obvious now as in the 1950s, it is still possible to show associations between day-to-day changes in air pollution and numbers of deaths and hospital admissions in those with heart and lung disease, as will be described in Chapter 2. So, it is likely that human health would benefit from further reductions in pollution, although action to bring this about could have other positive or negative consequences.

The United Kingdom (UK) National Air Quality Strategy and economic appraisal

1.3 The UK National Air Quality Strategy, published in March 1997, sets out goals for further reductions in the levels of air pollutants by the year 2005 (Department of the Environment, 1997) (see Table 1.1). The extent of the planned reductions was determined as follows. The first step was to set a series of air quality standards - defined levels of individual air pollutants intended to avoid or minimise significant risks to health. The next step was to consider whether it was actually technically or economically feasible to achieve the required reductions in pollution to these ideal levels. This resulted in a series of objectives. For some pollutants, it was reasonable to expect that the standard could be achieved so the objective was the same as the air quality standard. For others, the objectives were provisional and gave a percentage of occasions when the standard had to be met. This acknowledged that achieving the standard all the time could be so difficult that the disadvantages would outweigh the advantages (although there were many uncertainties in defining this balance precisely).

1.4 The Strategy states that it is a fundamental principle of Government policy that measures which incur a cost should achieve equivalent or greater benefits and that the option taken could not be substituted by another which achieves the same benefit at less cost. Although health benefits were assessed when setting the standards and the costs of compliance were considered when setting the objectives, the objectives in the Strategy were not based on explicitly comparing quantified estimates of costs and benefits - due to the uncertainties involved. However, the Strategy document does state that, in the first review of the strategy, the provisional objectives will receive special attention, particularly in relation to the costs and benefits of alternative measures and their relationship to the objectives. This report will contribute to the assessment of the benefits and will feed into the review of the Strategy now due to be completed by the end of 1998.

---

[1] This is the reason for the 'polluter pays' principle.

Table 1.1   Summary of Proposed Objectives in UK National Air Quality Strategy 1997

Pollutant	Standard	Objective to be achieved by 2005
Benzene	5 ppb running annual mean	5 ppb
1,3-Butadiene	1 ppb running annual mean	1 ppb
Carbon monoxide	10 ppm running 8-hour mean	10 ppm
Lead	0.5 $\mu g/m^3$ annual mean	0.5 $\mu g/m^3$
Nitrogen dioxide	150 ppb 1-hour mean 21 ppb annual mean	150 ppb 1-hour mean* 21 ppb annual mean*
Ozone	50 ppb running 8-hour mean	50 ppb measured as the 99th percentile*
Fine particles (PM_{10})	50 $\mu g/m^3$ running 24-hour mean	50 $\mu g/m^3$ measured as the 99th percentile*
Sulphur dioxide	100 ppb 15-minute mean	100 ppb measured as the 99.9th percentile*

ppm = parts per million;
ppb = parts per billion;
$\mu g/m^3$ = micrograms per cubic metre
* = these objectives are to be regarded as provisional

Terms of reference

1.5   This report has been prepared by an Ad-Hoc Group on the Economic Appraisal of the Health Effects of Air Pollution. This Group was set up by the Department of Health with the following terms of reference:

At the request of the Department of Health:

a) to advise on how best to reflect the importance of health effects in any cost/benefit decisions in air quality policy;

b) to estimate the healthcare cost implications of changes in levels of air pollution;

c) to consider whether monetary valuation of health effects is appropriate in this context;

d) to consider the merits of alternative approaches and, if necessary, to recommend further work to help develop the most appropriate approach;

e) if possible, to produce estimates of costs (with gaps in information and assumptions clearly stated) and, if necessary, to recommend further work to improve these estimates;

f) to produce a report within 12 months.

The group's work will be completed on production of the above report.

1.6 The membership of the Group includes experts in economic valuation of safety measures, environmental economics, health economics, risk analysis and the health effects of air pollution. The members of the Group are listed at Appendix 2. The Group has considered a range of relevant papers including discussion papers prepared by group members to develop ideas for the type of approach needed in this area. The Group has also been assisted by external advisers in particular areas such as epidemiology and their contribution is acknowledged in Appendix 2. The outcomes of these considerations have been drawn together in this report.

Overall approach

1.7 In interpreting the terms of reference, we have discussed several points of relevance to our overall approach. We wish to make clear that the report, and indeed other work on assessing costs and benefits, is intended only as a contribution to decision-making. Although important, quantified comparisons of costs and benefits provide only part of the information needed by decision-makers - some costs and benefits will remain impossible to quantify and there will also be distributional considerations to take into account. In addition, there are many uncertainties in analysing advantages and disadvantages of policies so it is unwise to expect a simple precise answer. We believe it is important that any assumptions and uncertainties involved are clearly stated.

1.8 The terms of reference cover only the human health effects of air pollution. This report does not, therefore, cover other benefits of reductions in air pollution such as reduced impacts on ecosystems, damage to buildings, etc. Neither does it cover the costs of implementing the policies required to reduce air pollution. However, the reduction in health effects is likely to make a significant contribution to the overall benefits of reductions in air pollution. Quantifying these health effects is a key step in assessing their importance relative to other consequences of reductions in air pollution. We have been able to draw on a recent report by the Committee on the Medical Effects of Air Pollutants on "Quantification of the Effects of Air Pollution on Health in the United Kingdom" (Department of Health, 1998). The findings of that report on the nature and numbers of adverse health events related to changes in air pollution are discussed in Chapter 2.

1.9 To aid the development of an appropriate policy on reducing air pollution, the health benefits need to be compared with other possible effects of the policy such as increased costs to industry or increased inconvenience to car drivers. There are many ways of doing this, for example, simply stating the likely health effects, using a scoring system or asking people how much they would be willing to pay for a small reduction in the risk to their health from air pollution. We have adopted an eclectic approach, considering the possiblities of using a range of different techniques and approaches. We have tried to avoid too many technical terms in describing these approaches but have provided a glossary at the end of the report (see Appendix 1) to assist in the understanding of any technical terms we have used. These approaches are considered further in Chapter 3.

1.10 The terms of reference specifically require the Group to examine the appropriateness of monetary valuation of health benefits. For a full cost-benefit analysis, the costs and benefits need to be in the same units for a direct comparison. Monetary valuation of the benefits thus allows direct comparison with the costs. However, we are aware that monetary valuation of benefits can be perceived with unease especially when applied to effects on health. (Also, monetary valuation of certain outcomes may not be feasible, or extremely difficult.) Discussion of this approach is, therefore, given particular attention in Chapter 3 and in the context of air pollution in Chapters 4 and 6.

1.11 The Group was set up to contribute to the assessment of the costs and benefits of meeting the objectives in the air quality strategy. In addition, we wished to ensure that our work was as useful as possible in other contexts. For example, the costs of health interventions are often compared with the gains in quality and duration of life. These gains are addressed in Chapters 3, 4, and 6 and health care costs are addressed in Chapter 5. We are aware that the benefits of reductions in air pollution are also being considered in deriving EC limit values for all pollutants (Council of the European Communities, 1998) and in work for the UNECE (AEA Technology, 1998; EFTEC, 1996). We hope this report will also be helpful in these contexts. This work should also be a contribution to the analysis of sustainable development.

1.12 This is primarily a methodological report. It aims to give advice on the range and appropriateness of the techniques available to represent the importance of the health benefits. If possible, valuation or other estimates per unit of air pollutant reduction are given. Provided the uncertainties are acknowledged, these can then be used in conjunction with projected pollution reductions for a more informed assessment of whether proposed measures are justified by the extent of their benefits. We wish to emphasise that this work is very much a first step in a new area but hope that one contribution of this report will be that our recommendations for further work in Chapter 7 will aid the development of better assessments in the future.

References

AEA Technology, Eyre Energy Environment, Metroeconomica (1998). *Cost Benefit Analysis of Proposals under the UNECE Multi-Pollutant, Multi-Effect Protocol. Final Report for the Department of the Environment, Transport and the Regions*. Harwell: AEA Technology.

Council of the European Communities (1998). Proposal for a Council Directive relating to limit values for sulphur dioxide, oxides of nitrogen, particulate matter and lead in ambient air (COM(97) 500 Final). *Off. J. Eur. Communities* **41**:C9/6-C9/20.

Department of the Environment (1997). *The United Kingdom National Air Quality Strategy*. London: The Stationery Office.

Department of Health (1998). Committee on the Medical Effects of Air Pollutants. *Quantification of the Effects of Air Pollution on Health in the United Kingdom*. London: The Stationery Office.

Ministry of Health (1954). *Reports on Public Health and Medical Subjects No. 95. Mortality and Morbidity During the London Fog of December 1952*. London: HMSO.

Research into Damage Valuation Estimates for Nitrogen Based Pollutants, Heavy Metals and Persistent Organic Pollutants. EFTEC, August 1996.

Chapter 2

Quantification of the Health Effects of Air Pollution

Previous reports

2.1 As mentioned in the Introduction, we have been able to identify the main health effects of reductions in air pollution by drawing on previous work by the Committee on the Medical Effects of Air Pollutants (COMEAP), in particular its most recent report "Quantification of the Effects of Air Pollution on Health in the United Kingdom" (Department of Health, 1998). This report assessed the available evidence on the health effects of air pollution and identified dose-response functions which could be applied with reasonable confidence in the UK. The criteria used in that report to judge whether the dose-response functions were suitable for use in quantification included the adequacy of control for confounding factors, the numbers of studies showing similar results and the likely transferability of studies from other countries to the UK.

Pollutants covered in this report

2.2 The COMEAP quantification report considered 5 pollutants - particulate matter less than 10 μm in diameter (PM_{10}), sulphur dioxide, ozone, nitrogen dioxide and carbon monoxide - and was able to identify dose-response functions suitable for quantification for PM_{10}, sulphur dioxide and ozone. Evidence for effects of nitrogen dioxide and carbon monoxide on health was presented but was not considered sufficiently robust for quantification. Nonetheless, we considered that some of the effects of these pollutants could be examined in a sensitivity analysis. Therefore, we have considered the spread of possible health outcomes resulting from these 5 pollutants in this report.

2.3 The remaining pollutants in the UK National Air Quality Strategy (Department of Environment, 1997) (benzene, 1,3-butadiene and lead) were not included in the COMEAP quantification report and will not be covered in detail here. However, we note that the characteristics of the potential health outcomes resulting from exposure to these 3 pollutants (e.g., development of leukaemia many years after exposure for benzene (Department of Environment, 1994a), development of lymphomas and leukaemias many years after exposure for 1,3-butadiene (Department of Environment, 1994b) and reduction in population IQ for lead (Department of Environment, Transport and the Regions, 1998)) differ in nature from the other health outcomes covered in the report.

Dose-response functions for use in quantification

2.4 The dose-response functions identified in the COMEAP report as suitable for use in quantification are given in Table 2.1.

2.5 It will be seen from Table 2.1, that just two types of health outcome were used for quantification - increases in all-cause mortality and increases in respiratory hospital admissions. It is important to understand the nature of these effects when reflecting their significance in economic appraisal. The COMEAP report notes that these dose-response functions are based on the results of time-series studies. These studies examine the relationship between daily levels of pollution and the risk of adverse health effects, on the same day or subsequent days, adjusting for climate and other factors. In the case of increases in all-cause mortality, all that is known is that, on average, the numbers of deaths increase approximately in proportion with the daily level of pollution. Little is known directly about the individuals who are dying since only aggregated statistics are used. In particular, it is not known whether exposure to air pollution means their deaths are brought forward by just a few days or by months or years. It is known that the increases in all-cause mortality are largely due to increases in respiratory and cardiovascular mortality and that the increases in all-cause mortality are higher in the elderly (probably because lung and heart disease are more common in this age group). Clinical judgement suggests that air pollution is acting as an aggravating factor in those who already have advanced lung or heart disease.

Table 2.1   **Exposure-response coefficients**

Pollutant	Health Outcome	Dose-Response Relationship
PM_{10}	Deaths brought forward (all causes)	+0.75% per 10 $\mu g/m^3$ (24 hour mean)
	Respiratory hospital admissions	+0.80% per 10 $\mu g/m^3$ (24 hour mean)
Sulphur dioxide	Deaths brought forward (all causes)	+0.6% per 10 $\mu g/m^3$ (24 hour mean)
	Respiratory hospital admissions	+0.5% per 10 $\mu g/m^3$ (24 hour mean)
Ozone	Deaths brought forward (all causes)	+0.6% per 10 $\mu g/m^3$ (8 hr mean)
	Respiratory hospital admissions	+0.7% per 10 $\mu g/m^3$ (8 hr mean)

2.6   The data relating levels of air pollution on particular days to respiratory hospital admissions are again based on aggregated statistics. Thus, it is not clear how many people are being admitted to hospital sooner than expected or how many people are being admitted to hospital who would not otherwise have been admitted to hospital at all. In addition, the studies do not distinguish between readmissions and first admission, i.e., the results could represent fewer people readmitted several times or more people admitted only once each.

Shape of dose-response relationships

2.7   The COMEAP report discusses the shape of the dose-response relationships, noting that, for the dose-ranges considered, it is reasonable to assume a linear relationship. In addition, for particles and sulphur dioxide, there was no good evidence of a threshold at the population level. This was not to say that there was not a threshold for individuals. However, because the concentration of an air pollutant at a fixed-site monitor acts only as a single estimate of an actual range of exposures and because there is likely to be a range of individual susceptibilities, demonstration of a threshold for the population would be unlikely. For ozone, data from London when plotted without assuming a linear model suggested a threshold at between 40 and 60 ppb (80-120 $\mu g/m^3$) but this was not supported by other studies. Calculations for ozone were, therefore, performed with and without a threshold assumption.

Uncertainty in dose-response coefficients

2.8   There are, of course, uncertainties in these estimated dose-response functions and the biological mechanisms may not be fully understood. Epidemiological studies demonstrate associations which may not necessarily be due to cause and effect. For example, levels of a pollutant may be correlated with a health effect because the pollutant rises and falls at the same time as another pollutant which is actually responsible. These uncertainties are lessened if there is explicit consideration of the effect of other pollutants and if many different studies report similar results. The COMEAP report considered these issues in detail and accepted that the associations used for quantification were likely to be causal. More generally, the issue of uncertainty is very important because, in a complex multi-stage process such as quantifying and valuing health effects, the level of uncertainty is likely to accumulate at each stage.

2.9 The COMEAP report also considered the problem of providing some indication of the likely accuracy of the estimated dose-response functions. The dose-response functions used were based on meta-analyses by either the World Health Organisation (WHO) or a European project on air pollution (APHEA). These used a selected range of studies for good reasons (for example, the APHEA study is a meta-analysis of a series of European studies with a matching protocol) but they did not cover the full range of studies in the literature. The COMEAP report did not consider it was appropriate to calculate formal confidence limits since the Committee had selected the APHEA and WHO results as "best estimates" rather than doing a fully comprehensive meta-analysis. The report also stated COMEAP's concern that informal lower and upper bounds might be liable to misinterpretation. This is because the confidence limits only reflect sampling errors and not other uncertainties. However, we consider that, provided it is clear that not all possible studies were covered, the confidence limits for the WHO and APHEA analyses would allow some aspects of the uncertainty to be reflected in the sensitivity analysis. The relevant ranges are given in Annex 2A.

Dose-response functions for use in sensitivity analysis

2.10 The COMEAP Quantification Report used carefully specified criteria in determining which dose-response functions were sufficiently robust for use in quantifying the health effects of air pollutants. For example, studies needed to be of good quality with adequate consideration of confounding factors. More weight was given to outcomes for which a range of studies had shown similar effects. Preference was given to UK or European studies (many of which included UK data) when selecting dose-response functions as these were more likely to be applicable to the UK than studies from other countries around the world. The inclusion or exclusion of particular dose-response functions could, obviously, have a significant effect on the overall assessment of the benefits. We consider it is important to examine the effect that different dose-response functions might have. In addition to the dose-response functions used for quantification, the COMEAP report describes the overall evidence on the health effects of each pollutant including a range of other dose-response functions. The quality of evidence behind these dose-response functions varies from single high quality studies which may well soon be confirmed by other studies, to a range of inconsistent studies with poorly defined endpoints.

2.11 We therefore consulted some of the experts involved in drawing up the COMEAP report on which of these other dose-response functions were suitable for use in sensitivity analysis[1]. The text below is based on this advice. It is important that the dose-response functions are used for sensitivity analysis only and the associated uncertainties are always quoted with the dose-response function. We consider some more general issues of relevance to the sensitivity analysis before considering dose-response functions for sensitivity analysis on different pollutants and outcomes. The intention here is to give a broad indication of the sort of range in which the dose-response function might fall. Therefore, the studies are not discussed in detail (most are, in any case, covered in the COMEAP quantification report) and rigid criteria for selection of an appropriate function are not required.

---

[1] Sensitivity analysis - the testing of the "sensitivity" of the final answer to variation in important inputs (such as dose-response functions) to the calculations. (See glossary).

General issues

Life-expectancy

2.12 We noted in paragraph 2.5 that it was unknown whether the acute effects of air pollution brought deaths forward by days, months or even years. Although direct evidence is lacking, judgement based on knowledge of patients with respiratory disease, estimates of loss of life expectancy in long-term studies of the effects of air pollution (Brunekreef, 1997; Miller and Hurley, 1998) and estimates of loss of life expectancy due to other more serious factors such as cigarette smoking (Peto *et al*, 1996) suggest a loss of life expectancy measured in days, weeks or months is more likely than a loss measured in years. We suggest that examples of possible average losses of life expectancy for respiratory deaths of 1 month, 6 months and a year could be examined when performing a sensitivity analysis but it should be noted that average loss of life expectancy might be less than 1 month.

Age

2.13 The hospital admissions associated with air pollution are due to cardio-respiratory diseases and occur predominantly in the elderly. Some studies have derived dose-response functions separately for those under and over 65 (Spix *et al*, 1998). Use of specific dose-response functions for those over and under 65 rather than for all ages may affect the overall assessment as there may be parts of the country with larger numbers of elderly people than other areas. If the areas with more elderly people happen to be areas with higher levels of pollution then the total impact will be increased. Conversely, if, for example, the elderly retire to less polluted areas the total impact will be decreased. In addition, disaggregation by age could have an effect on later stages of the economic appraisal.

2.14 The COMEAP report quantified all-cause mortality and respiratory hospital admissions for all ages. For all cause mortality, the report does not specifically discuss dose-response functions for different age groups. Relatively few studies of all cause mortality appear to have examined dose-response functions for over and under 65s separately. In Annex 2A, dose-response functions for particles from one such study in the US (Schwartz and Dockery, 1992) are suggested while noting that functions in Europe (Verhoeff *et al*, 1996) might be lower. However, the COMEAP report does quote respiratory hospital admission dose-response functions for the over and under 65s from the APHEA meta-analysis (Spix *et al*, 1998). We suggest that the over and under 65 dose-response functions for respiratory admissions for sulphur dioxide and/or ozone are used to test in a sensitivity analysis whether disaggregation by age is likely to have a significant effect on the overall assessment of the benefits. Small age differences have also been shown for cardiovascular admissions (Burnett *et al*, 1995) but, given the fact that this endpoint was not used for quantification, we do not suggest a sensitivity analysis on age effects for this endpoint.

Thresholds

2.15 As mentioned above in para 2.7, the COMEAP report did calculations with and without an assumed threshold of 50 ppb for the effect of ozone on respiratory hospital admissions and deaths brought forward. This had a significant effect on the results. We suggest the presence or absence of a 50 ppb threshold is also used in sensitivity analyses of other work quantifying benefits.

Inclusion or not of further pollutants

2.16 As noted in paragraph 2.2, the COMEAP report did not consider the evidence on the health effects of nitrogen dioxide and carbon monoxide sufficiently robust for quantification. However, there is some indication that these pollutants do have some effect on health. How much does the exclusion of these pollutants and their possible effects matter?

Nitrogen dioxide (NO_2)

2.17 The COMEAP report records some of the inconsistencies in the evidence for effects of nitrogen dioxide. Increases in all cause mortality have been shown (Touloumi *et al*, 1997) but this was not explained by increases in cardiovascular or respiratory mortality as would be expected (Zmirou *et al*, 1998). This is puzzling and we suggest the results are not used in sensitivity analyses (the effect estimate is in any case lower than that for other pollutants). Increases in total respiratory admissions were not shown (Spix *et al*, 1998) but increases in asthma admissions (Sunyer *et al*, 1997) and admissions for chronic obstructive pulmonary disease were found (Anderson *et al*, 1997). These were used to estimate a dose response function for respiratory admissions (see Annex 2A), although this was not used for quantification. However, this function could be used for sensitivity analysis to test the implications of including effects of nitrogen dioxide in the assessment of benefits. As nitrogen dioxide and particles are both generated by traffic they are often present together. Control for particles has not always been adequate in studies which report an effect of nitrogen dioxide. The difficulty in separating the effects of nitrogen dioxide from the effect of particles should be emphasised when performing such a sensitivity analysis.

Carbon Monoxide (CO)

2.18 The COMEAP report also describes the evidence for effects of carbon monoxide on all-cause and cardiovascular mortality and on cardiovascular admissions. The effects of CO on the oxygen carrying capacity of haemoglobin and consequent effects on the heart are well known (Institute for Environment and Health, 1998). However, effects on cardiovascular outcomes at the low levels of CO present in the environment are less well established. COMEAP noted the lack of UK studies (only one recent study in London was available (Poloniecki *et al*, 1997)), uncertainties regarding the effect of CO alone as compared with that of the urban pollution mixture and the fact that fixed site monitors may reflect average population exposure to CO less well than for other pollutants (Cortese and Spengler, 1976). It was considered that the evidence could not be used for quantification. However, it was recognised that evidence was accumulating rapidly and assessment might be possible soon. Some dose-response functions for cardiovascular admissions and all-cause mortality which could be used for sensitivity analyses are described in Annex 2A. Because the studies are relatively recent, the mechanism of effect at low levels is unclear and there may be difficulties regarding the exposure estimates, it may currently be unwise to use these functions even for a sensitivity analysis.

Inclusion or not of health outcomes other than all cause mortality and respiratory hospital admissions

2.19 The COMEAP report quantified only all cause mortality and respiratory hospital admissions but did describe dose-response functions for other health outcomes. This section discusses which of these are suitable for use in sensitivity analysis.

Cause-specific mortality

2.20 The COMEAP report considered whether to quantify the effects of air pollutants on respiratory mortality and cardiovascular mortality but did not do so. For example, for ozone, it was considered that there were too few studies on cause-specific mortality. Dose-response functions for all-cause mortality represent the effects of pollutants on mortality without the potential difficulties with variations in diagnosis of cause of death. (If more studies were available, cause-specific mortality could be useful particularly where the baseline cause-specific mortality varied across areas.) There could be an argument for using cause-specific mortality if there were indications that people valued reductions in risks of cardiovascular deaths differently from reductions in risks of respiratory deaths. (For example, losses of life expectancy might be greater for cardiovascular deaths). On current evidence, particles, sulphur dioxide and ozone all cause slightly lower increases in deaths brought forward from cardiovascular disease than from respiratory disease (Zmirou *et al*, 1998). However, irrespective of air pollution, there are 1.8 times more cardiovascular deaths than respiratory deaths (Office for National Statistics, 1998) so overall the numbers of cardiovascular and respiratory deaths may be similar. It would be useful to explore this further when more evidence is available. However, at this stage, for simplicity, we do not suggest the use of cause-specific mortality in a sensitivity analysis.

Cardiovascular Admissions

2.21 Associations have been shown between cardiovascular admissions and increased levels of several pollutants. The effect of carbon monoxide on cardiovascular admissions has been discussed above in paragraph 2.18. The COMEAP report also describes the evidence for an effect of particles on cardiovascular admissions and concludes that, although the overall evidence does suggest an effect, the number of studies are relatively few and there are important differences between studies regarding which types of heart disease are involved and the estimated sizes of the effects. For these reasons the evidence was not used for quantification.

2.22 The effect of air pollution on heart disease is an active area of research and could potentially be a significant contributor to the overall health effects. We have already suggested that the health effects of carbon monoxide, including cardiovascular admissions, should be included in a sensitivity analysis. We suggest the addition of cardiovascular admissions to the health effects of particles should also be examined in a sensitivity analysis. A possible dose-response function is discussed in Annex 2A.

Respiratory symptoms

2.23 We are aware that air pollution also has an effect on more minor symptoms and that this could potentially affect a larger number of people than deaths or hospital admissions. The COMEAP report describes a range of studies which have examined particular groups (often with known respiratory disease) for changes in respiratory symptoms or bronchodilator usage. These outcomes overlap to some extent and subjects may vary in what they define as an increase in respiratory symptoms. Possible sources for dose-response functions are noted in Annex 2A but the added problems of defining the size of the appropriate subject population (asthmatics variously defined, children with chronic respiratory symptoms, etc.) and the appropriate baseline rates may mean a sensitivity analysis cannot in any case be fully pursued.

2.24 We are aware of a study in London examining associations between GP consultations and various air pollutants which is not yet published. This will add to the understanding of the effect of air pollution on less serious episodes of disease.

Mortality due to chronic exposure to air pollution

2.25 In 1995, COMEAP advised that, although the evidence for the chronic effects of particles was limited, it would be prudent to consider the associations as causal (Department of Health, 1995). The 1998 COMEAP quantification report did not consider that the evidence for chronic effects of air pollution was sufficiently robust to use for quantification although it was noted that, for chronic exposure to particles, the impact on mortality could be substantial. There was evidence from cohort studies in the US (Dockery *et al*, 1993; Pope *et al*, 1995) which offered support for chronic effects of air pollution, particularly for fine particles, but an absence of suitable UK or European cohort or longitudinal studies. The report noted the uncertainties in estimating measures of the impact of chronic exposure such as years of life lost.

2.26 Work in progress in the UK (Miller and Hurley, 1998) has been examining the potential impact of the chronic effects of particles using a range of assumptions (several of which are currently untestable). For example, the published papers do not give information on the lag time between exposure to a particular level of particles and death (which might in any case vary from person to person). Since the effects on acute mortality would contribute to the differences in life expectancy estimated from the cohort studies, at least some of the effect would take place immediately. In addition, the form of analysis in the cohort studies assumes that the proportional effect on mortality hazards of long-term exposure to a given concentration of ambient particles is the same at different ages. This is not necessarily the case if, for example, hazards increase with increasing years of exposure to pollutants. However, it is possible to choose some particular assumptions and, using lifetables and a computer program developed by the UK project (Miller and Hurley, 1998), to compare two scenarios with and without a reduction in pollution. Results can be expressed in terms of years of life lost or numbers of deaths in a particular age group in a particular year.

2.27 The Group has had limited time to give full consideration to this complex issue. We will not discuss it further here apart from noting that the effects on mortality predicted by the cohort studies cannot simply be added to the acute effects on mortality predicted by the time-series studies because the cohort studies include both acute and chronic effects. It is important that work continues on estimating the possible public health impact of mortality due to chronic exposure given the potential size of the effect.

Chronic morbidity

2.28 We do not consider the evidence for effects on chronic morbidity is sufficiently well founded for use in sensitivity analysis. The number of studies are limited and there are some difficulties with inconsistency and confounding. The studies that are available and the difficulties in interpretation are discussed in the COMEAP report (Department of Health, 1998).

Other outcomes

2.29 Some outcomes for which we do not consider suitable dose-response functions can be identified for use in sensitivity analysis have already been discussed above. In addition, we consider that "emergency room visits" is an endpoint specific to the US healthcare system (more patients with milder disease attend than would be the case in the UK where patients would go to their GP instead) and that "restricted activity days" are too vaguely and subjectively defined to be meaningful for quantification.

Outputs needed for quantification and for cost-benefit analysis

2.30 As discussed in the Introduction, this report is intended to feed into a cost-benefit analysis of policies to meet the objectives in the National Air Quality Strategy. As a primarily methodological report, this report will not be quantifying, valuing or aggregating the benefits of the required reductions in pollutants. This will be part of the full cost-benefit analysis. Nonetheless, it is important to be aware of the procedures which will be used in this cost-benefit analysis so that we can be sure that the outputs from our report are presented in a form that can be used in the relevant calculations.

2.31 The National Air Quality Strategy aims to meet specified objectives for each pollutant by the year 2005. Work is being undertaken as part of the review of the Strategy to predict the likely concentration of each pollutant in 2005 on the basis of current policies. If this predicted concentration does not meet the objective, additional policies may be needed. To assist in the consideration of whether to recommend these additional policies and the consideration of the optimal level for the objectives, the costs and benefits of the available policies will be examined.

2.32 Predicted pollutant concentrations for 2005 for each 1km grid square across the country will be available and the difference between this and the desired concentration will be used to calculate the benefits to health. The following inputs are needed for this calculation[2].

(i) numbers of people in the grid square in the relevant population group (general population, elderly, asthmatics etc.);

(ii) the baseline rate of the appropriate health outcome in that population (in numbers of health outcomes per standard number of people in the relevant population group);

(iii) the dose-response function for the appropriate health outcome (percentage increase in baseline numbers of health outcomes per unit change in concentration of pollutant);

(iv) change in pollutant concentration needed in that grid square to meet objective or to be generated by a particular policy measure under consideration.

[2] Strictly, the change in health outcome for a different increment in pollutant concentration should be calculated using the formula: ln (Relative Risk) = ß x increment in pollutant concentration where ß is the regression coefficient from the time-series studies. However, for the low relative risks used here we found this made little practical difference to the answer. In addition, the calculation should take account of the fact that air pollution may already be contributing to the baseline level of the relevant health outcome. However, in practice, the numbers of deaths brought forward or hospital admissions influenced by air pollution is so small relative to the total numbers of deaths and hospital admissions that the simpler calculation provides a reasonable approximation.

Multiplying these together gives the expected reduction in numbers of health events (e.g., hospital admissions) for that grid square. These can then be summed for all grid squares.

2.33 The dose-response functions and the population groups to which they relate have been presented earlier in this chapter. In many cases, identification of the population of the relevant group can be obtained from Census data. However, it is less straightforward for particular disease groups since there can be problems of disease definition and the studies used to generate the dose-response function may have used their own case definitions. Statistics on baseline rates are published (Office for National Statistics, 1998; Department of Health, 1996). These should match those used in the studies generating the dose-response functions. For example, most studies of air pollution and mortality exclude accidents (ICD Code >800). Studies of hospital admissions are often restricted to emergency admissions although this is not possible in all places. The majority of the studies contributing to the APHEA meta-analysis used deaths excluding external causes and emergency hospital admissions only.

2.34 We do not have the information on pollutant concentrations specified in 2.32(iv) above. However, we can calculate the numbers of health outcomes per unit change in concentration of pollutant. In the COMEAP report, the calculations for particles and sulphur dioxide were based on the population in urban areas of Great Britain (full background data was not then available for Northern Ireland). This was because the epidemiological studies are done in cities. For ozone, the calculations were based on the population of both urban and rural areas of Great Britain but were performed for the summer only. Ozone is generated by sunlight and is much higher in summer. It is also higher in rural areas. The figures below are on the same basis although, for simplicity, we have done the calculations at the level of the total population rather than summing calculations by grid square. This will be done in the cost-benefit analysis.

2.35 We have used the dose-response functions in Table 2.1 (with the ranges in uncertainty from Annex 2A) and the following baseline data:

For particles and sulphur dioxide:

GB Urban population 42.5 million
1074 deaths excluding external causes per 100,000 people all year (1995)
Emergency respiratory admissions 830 per 100,000 people all year (1994/5)

For ozone:

GB Urban plus rural population 57 million
491.8 deaths excluding external causes per 100,000 per summer (1995)
Emergency respiratory admissions 360 per 100,000 summer (April to September 1995).

For example, the numbers of deaths excluding external causes in the GB urban population per year is 1074/100,000 x 42.5million i.e., about 455,000. Particles increase deaths brought forward by 0.75% per 10 $\mu g/m^3$ (0.075% per $\mu g/m^3$) so the numbers of deaths brought forward are 0.075/100 x 455,000 or about 340. The results for particles, sulphur dioxide and ozone are given in Table 2.2 below.

These numbers are used in later chapters and are discussed further in the conclusions chapter.

Table 2.2 Summary of reductions in numbers of deaths brought forward and respiratory hospital admissions per $\mu g/m^3$ reduction in pollutant per year or per summer

	Health outcome per $\mu g/m^3$ reduction of pollutant[a]		
	PM_{10} in GB urban population per year	Sulphur dioxide[b] in GB urban population per year	Ozone[c] in GB urban and rural population summer only
Reduction in deaths brought forward (all cause) (acute)	340 (270 - 385)	270 (225-315)	170 (70-225)
Reduction in respiratory hospital admissions (emergency)d	280 (180-390)	180 (-53 to 320)	145 (50-245)

Notes to Table 2.2:

[a] These estimates would increase if the Northern Ireland population were included as will be done in the full cost-benefit analysis.
[b] Statistically the reduction in hospital admissions of -53 represents a small possibility of a beneficial effect of sulphur dioxide but other evidence such as adverse effects of sulphur dioxide on volunteers in chamber studies does not support this.
[c] The estimated effect of ozone on hospital admissions will be lower if there is a threshold at about 50 ppb. This calculation requires information on predicted days above and below the threshold. This information was not available to the Group.
[d] There has been a small change in the calculation of baseline emergency respiratory admission rates since completion of these calculations. This would have a minimal effect (an increase of about 0.8%).

2.36 Once the health outcomes have been quantified they need to be expressed in a way that allows them to be compared with the costs of implementing the additional policies. The following chapters address how this might be done.

References

Anderson, H.R., Spix, C., Medina, S., Schouten, J., Castellsague, J., Rossi, G., Zmirou, D., Touloumi, G., Wojtyniak, B., Pönkä, A., Bachárová, L., Schwartz, J. and Katsouyanni, K. (1997) Air pollution and daily admissions for chronic obstructive pulmonary disease in 6 European cities: results from the APHEA project. *Eur. Respir. J.* **10**: 1064-1071.

Brunekreef, B. (1997) Air pollution and life expectancy: is there a relation? *Occup. Environ. Med.* **54**: 731-734.

Burnett, R.T., Dales, R.E., Krewski, D., Vincent, R., Dann, T. and Brook, J. (1995) Associations between ambient particulate sulfate and admissions to Ontario hospitals for cardiac and respiratory diseases. *Am. J. Epidemiol.* **142**: 15-22.

Cortese, A.D. and Spengler, J.D. (1976) Ability of fixed monitoring stations to represent personal carbon monoxide exposure. *J. Air Pollut. Control Assoc.* **26**: 1144-1150.

Department of Health (1996). *Hospital Episode Statistics. Volume 1. Finished Consultant Episodes by Diagnosis and Operative Procedure; Injury/Poisoning by External Causes.* London: Department of Health.

Department of Health (1995). Committee on the Medical Effects of Air Pollutants. *Non-Biological Particles and Health.* London: The Stationery Office.

Department of Health (1998). Committee on the Medical Effects of Air Pollutants. *Quantification of the Effects of Air Pollution on Health in the United Kingdom.* London: The Stationery Office.

Department of the Environment (1997). *The United Kingdom National Air Quality Strategy.* London: The Stationery Office.

Department of the Environment (1994a). Expert Panel on Air Quality Standards. *Benzene.* London: HMSO.

Department of the Environment (1994b). Expert Panel on Air Quality Standards. *1,3-Butadiene.* London: HMSO.

Department of the Environment, Transport and the Regions (1998). Expert Panel on Air Quality Standards. *Lead.* London: The Stationery Office.

Dockery, D.W., Pope, C.A., Xu, X., Spengler, J.D., Ware, J., Fay, M.A., Ferris, B.G. and Speizer, F.E. (1993) An association between air pollution and mortality in six US cities. *N. Engl. J. Med.* **329**: 1753-1759.

Institute for Environment and Health (1998). *IEH Assessment on Indoor Air Quality in the Home (2): Carbon Monoxide.* Leicester: Institute for Environment and Health.

Miller, B. and Hurley, F. (1998) *Towards Quantifying and Costing the Overall Effects of Ambient Air Pollution on Mortality in the UK.* Edinburgh: Institute of Occupational Medicine.

Office for National Statistics (1998). *1996 Mortality Statistics. Cause. Review of the Registrar General on Deaths by Cause, Sex and Age in England and Wales, 1996. Series DH2 No 23.* London: The Stationery Office.

Peto, R., Lopez, A.D., Boreham, J., Thun, M., Heath, C. and Doll, R. (1996) Mortality from smoking worldwide. *Br. Med. Bull.* **52**: 12-21.

Poloniecki, J.D., Atkinson, R.W., Ponce de Leon, A. and Anderson, H.R. (1997) Daily time series for cardiovascular hospital admissions and previous day's air pollution in London, UK. *Occup. Environ. Med.* **54**: 535-540.

Pope, C.A., Thun, M.J., Namboodiri, M.M., Dockery, D.W., Evans, J.S., Speizer, F.E., and Heath, C.W. (1995) Particulate air pollution as a predictor of mortality in a prospective study of US adults. *Am. J. Respir. Crit. Care Med.* **151**: 669-674.

Schwartz, J. and Dockery, D.W. (1992) Increased mortality in Philadelphia associated with daily air pollution concentrations. *Am. Rev. Respir. Dis.* **145**: 600-604.

Spix, C., Anderson, H.R., Schwartz, J., Vigotti, M.A., LeTertre, A., Vonk, J.M., Touloumi, G., Balducci, F., Piekarski, T., Bacharova, L., Tobias, A., Pönkä, A. and Katsouyanni, K. (1998) Short-term effects of air pollution on hospital admissions of respiratory diseases in Europe: a quantitative summary of APHEA study results. *Arch. Environ. Health* **53**: 54-64.

Sunyer, J., Spix, C., Quénel, P., Ponce de Leon, A., Pönkä, A., Barumandzadeh, T., Touloumi, G., Bachárová, L., Wojtyniak, B., Vonk, J., Bisanti, L., Schwartz, J. and Katsouyanni, K. (1997) Urban air pollution and emergency admissions for asthma in four European cities: the APHEA project. *Thorax* **52**: 760-765.

Touloumi, G., Katsouyanni, K., Zmirou, D., Schwartz, J., Spix, C., Ponce de Leon, A., Tobias, A., Quennel, P., Rabczenko, D., Bachárová, L., Bisanti, L., Vonk, J.M. and Pönkä, A. (1997) Short-term effects of ambient oxidant exposure on mortality: a combined analysis within the APHEA project. *Am. J. Epidemiol.* **146**: 177-185.

Verhoeff, A.P., Hoek, G., Schwartz, J. and van Wijnen, J.H. (1996) Air pollution and daily mortality in Amsterdam. *Epidemiology* **7**: 225-230.

Zmirou, D., Schwartz, J., Saez, M., Zanobetti, A., Wojtyniak, B., Touloumi, G., Spix, C., Ponce de Leon, A., Le Moullec, Y., Bachárová, L., Schouten, J., Pönkä, A. and Katsouyanni, K. (1998) Time-series analysis of air pollution and cause-specific mortality. *Epidemiology* **9**: 495-503.

ANNEX 2A

Suggested Dose-Response Functions for Sensitivity Analyses

1. The dose-response functions below are intended for use in sensitivity analyses *only*, to give a rough guide to the *possible* importance of the various outcomes. The COMEAP quantification report (Department of Health, 1998) only recommends specific overall dose-response functions for full quantification (see Table 2.1 in main text and Table 2A.1 below) but does cover other endpoints in its review of the literature. The suggestions below are based on that report and on discussions with COMEAP members. The uncertainties associated with each suggested function are stated and it is important that these are quoted whenever the functions are used. Since the functions are only intended to give a rough guide, they are selected as typical or worst case values from a range of studies but are not "best estimates" selected according to rigid criteria.

Table 2A.1   Exposure-response coefficients with rough range of uncertainty

Pollutant	Health Outcome	Dose-Response Relationship (rough uncertainty range)
PM_{10}	Deaths brought forward (all causes)	+0.75% per 10 $\mu g/m^3$ (24 hour mean) (+0.6 to +0.85) (WHO) (World Health Organisation, 1998)
	Respiratory hospital admissions	+0.80% per 10 $\mu g/m^3$ (24 hour mean) (+0.5 to +1.1) (WHO) (World Health Organisation, 1998)
Sulphur dioxide	Deaths brought forward (all causes)	+0.6% per 10 $\mu g/m^3$ (24 hour mean) (+0.5 to +0.7) (APHEA) (Spix *et al*, 1998)
	Respiratory hospital admissions	+0.5% per 10 $\mu g/m^3$ (24 hour mean) (-0.15 to +0.9)*(APHEA) (Spix *et al*, 1998)
Ozone	Deaths brought forward (all causes)	+0.6% per 10 $\mu g/m^3$ (8 hr mean) (+0.25 to +0.8) (APHEA) (Spix *et al*, 1998)
	Respiratory hospital admissions	+0.7% per 10 $\mu g/m^3$ (8 hr mean) (+0.25 to +1.2)* (APHEA) (Spix *et al*, 1998)

* Based on lowest confidence limit for 15-64 age group and highest confidence limit for 65+ age group.

Statistical Uncertainty Around Dose-Response Functions Used for Quantification

2   It was suggested in paragraph 2.9 that a rough range of uncertainty could be derived from the lowest and highest confidence intervals in the range of studies considered for the overall dose-response function (See Table 2A.1). It was noted that these were not formal confidence intervals since the dose-response functions had not necessarily been derived from a formal meta-analysis. These give some guide to uncertainty due to sampling errors but do not cover other aspects of uncertainty.

Age

All-cause mortality

3   Greater effects on mortality amongst the elderly were noted during the London smog in 1952 (Ministry of Health, 1954). This was also shown for particles by Schwartz and Dockery (1992) and in other more recent studies (United States Environmental Protection Agency, 1995). The APHEA study (Katsouyanni et al, 1997) did not consider all-cause mortality separately by age. Another study in Europe (Verhoeff et al, 1996) did show a slightly higher relative risk for the effects of particles in the elderly. Other pollutants do not appear to have been studied to the same extent. It is suggested that, for illustrative purposes, the dose-response functions for particles from Schwartz and Dockery (1992) are used but that it is acknowledged that the figures for Europe could be lower. The functions were 7%, 3% and 10% increases per 100 $\mu g/m^3$ TSP (total suspended particulates) for all ages, under 65 and over 65, respectively. Using a conversion factor of PM_{10} = TSP x 0.55, these equate to:

all ages   1.2% increase per 10 $\mu g/m^3$ PM_{10}

under 65   0.5 % increase per 10 $\mu g/m^3$ PM_{10}

over 65   1.8% increase per 10 $\mu g/m^3$ PM_{10}

4   These functions can be used to demonstrate the large proportion of deaths brought forward in the elderly. In Philadelphia, where the study was performed, 65% of all deaths were in those over 65. Applying the above increases suggests (1.8% x 65)/((1.8% x 65) + (0.5% x 35)) = 87% of deaths brought forward by particulate air pollution occur in the elderly. The numbers may not be exactly the same for the UK population but a similarly high proportion of the elderly would be expected to be affected.

5   This can also be used to demonstrate the difference in level of risk between the general population and the elderly. The COMEAP report estimates that, due to PM_{10}, 8100 deaths are brought forward in 1 year in the GB urban population of 42.5 million - a risk of 2 x 10^{-4}. If 87% of the 8100 deaths brought forward by particulate pollution are in the elderly (7050) and the elderly form about 16% of the population (Department of Health, 1998) (6.8 million for GB urban population), then the risk in the elderly would be about 7050/(6.8 x 10^6) = 1 x 10^3.

Respiratory hospital admissions

6   The effect of air pollution on respiratory hospital admissions was considered for the <65 and 65+ age groups in the APHEA study (Spix et al, 1998). As this was a meta-analysis of results from several European cities including London, this would be a reasonable study to use to derive dose-response functions for sensitivity analyses. Functions are given below for sulphur dioxide and ozone. For particles, the results were expressed using black smoke or TSP rather than PM_{10} and did not show a clear difference of effect by age group.

Sulphur dioxide (daily mean)	15-64 y	0.2% per 10 $\mu g/m^3$
	65+	0.4% per 10 $\mu g/m^3$
Ozone (8 hour average)	15-64 y	0.6% per 10 $\mu g/m^3$
	65+	0.75% per 10 $\mu g/m^3$

The functions for either or both these pollutants could be used to test the sensitivity of the calculations to disaggregation by age.

Nitrogen dioxide

7   In paragraph 2.17, it was suggested that a dose-response function for respiratory hospital admissions but not mortality should be considered for a sensitivity analysis of the inclusion (or not) of nitrogen dioxide. In the APHEA study no significant increase in overall respiratory hospital admissions was seen (Spix *et al*, 1998) but there was a 2.6% per 50 $\mu g/m^3$ increase in asthma admissions (Sunyer *et al*, 1997 and a 1.9% per 50 $\mu g/m^3$ increase in chronic obstructive pulmonary disease (COPD) admissions (Anderson *et al*, 1997). Based on these results, the COMEAP quantification report (Department of Health, 1998) suggested a dose-response function of a 2.5% increase in respiratory hospital admissions per 50 $\mu g/m^3$ (0.5% per 10 $\mu g/m^3$) but did not recommend that this was used for full quantification.

Carbon monoxide

8   Paragraph 2.18 notes some of the uncertainties behind the use of dose-response functions for carbon monoxide (only one UK study and difficulties in separation from other pollutants). There is no UK study examining the effect of carbon monoxide on all-cause mortality. A study in Athens (Touloumi *et al*, 1996) found an effect, but there was no discussion of the effect of other pollutants on the dose-response estimate. Other studies from the United States (Department of Health, 1998) have found an effect on all-cause mortality independent of other pollutants.

9   The effect of carbon monoxide on hospital admissions has been studied in London (Poloniecki *et al*, 1997), the United States (Schwartz and Morris, 1995; Morris *et al*, 1995; Schwartz, 1997) and Canada (Burnett *et al*, 1997). Two studies found an effect on acute myocardial infarction (Poloniecki *et al*, 1997) or ischaemic heart disease (Schwartz and Morris, 1995) but the effect did not remain after control for other pollutants. Three studies (Schwartz and Morris, 1995; Morris *et al*, 1995; Burnett *et al*, 1997) found effects on congestive heart failure which did remain after control for other pollutants but the London study (Poloniecki *et al*, 1997) found no significant effect. Effects on combined circulatory diseases (in only one of two statistical models) (Poloniecki *et al*, 1997) and on all cardiovascular admissions (independent of other pollutants) (Schwartz, 1997) have also been shown.

10 The COMEAP report (Department of Health, 1998) describes the dose-response functions from these studies and notes in summary that a rise of 10 ppm carbon monoxide is associated with increases of about 10% in all-cause mortality and about 20% in cardiovascular admissions. These could be used for sensitivity analyses but it should be noted that the averaging time for the CO measurement is unclear (the studies used the maximum 8 hour average (Touloumi *et al*, 1996), the 24 hour average (Poloniecki *et al*, 1997) and the maximum 1 hour average (Schwartz and Morris, 1995; Morris *et al*, 1995; Burnett *et al*, 1997)) and the uncertainty as to whether fixed site monitors provide a reasonable estimate of average population exposure is greater for carbon monoxide than for other pollutants (Cortese and Spengler, 1976).

Cardiovascular (and cerebrovascular) admissions

11 Studies of the effects of air pollutants on cardiovascular admissions are recent and relatively few in number. The studies vary as to which pollutants and which endpoints were examined. None of the studies showed any effect of ozone on cardiovascular admissions. Sulphur dioxide does not generally have an effect which persists in multi-pollutant models apart from an effect on acute myocardial infarction in the London study (Poloniecki *et al*, 1997). Nitrogen dioxide also does not generally have an effect which persists in multi-pollutant models although an effect on congestive heart failure was shown in Canada (Burnett *et al*, 1997). Evidence for an effect of carbon monoxide is better although still unclear in detail as discussed above.

12. The evidence for an effect of particles on cardiovascular admissions is better although still confusing as to the important endpoints. The use of different measures of particles such as black smoke (Poloniecki *et al*, 1997), coefficient of haze (Burnett *et al*, 1997) and sulphates (Burnett *et al*, 1995) also adds to the difficulties in interpretation. The UK studies (Poloniecki *et al*, 1997; Wordley *et al*, 1997) were for all ages, most others were for the elderly. Studies of acute myocardial infarction or ischaemic heart disease have found no effect (Wordley *et al*, 1997); an effect which was not independent of control for other pollutants (Poloniecki *et al*, 1997) and an effect which was independent of control for other pollutants (Schwartz and Morris, 1995). The picture is similar for congestive heart failure with evidence for no effect (Poloniecki *et al*, 1997), no independent effect (Burnett *et al*, 1997) and a clear independent effect (Schwartz and Morris, 1995). Effects on all cardiovascular admissions of PM_{10} (independent of other pollutants) (Schwartz, 1997) and of sulphates (Burnett *et al*, 1995) have also been shown. Effects on cerebrovascular diseases were found in Birmingham (Wordley *et al*, 1997) but not in London (Poloniecki *et al*, 1997). The latter study did find an effect on combined circulatory diseases but in only one of two statistical models.

13 Given the contradictions over the exact endpoint involved, the use of an overall dose-response function for cardiovascular admissions such as that in Schwartz (1997) (around 1% increase per 10 $\mu g/m^3$ PM_{10} for the elderly) is probably the most sensible. The possibility of an effect on cerebrovascular disease (which is not covered by cardiovascular admissions) should be noted.

Respiratory symptoms

14 Although an effect of air pollution on respiratory symptoms could potentially affect a large number of people, there are some difficulties in the use of any such evidence for predicting its public health impact. In contrast to the studies of mortality and hospital admissions, the studies generally examine small groups of people often with a particular disease. This can be an advantage for learning about effects on sensitive groups but means that data on the number of people in the population in that sensitive group and the baseline level of symptoms in that sensitive group will be needed to assess the overall impact. This is not always available. The studies are less likely to control for other pollutants and the results are difficult to summarise across a range of studies examining slightly different sub-groups.

15 The COMEAP report describes the effects of the various pollutants on respiratory symptoms and bronchodilator usage. Changes in bronchodilator usage are less subjective than changes in respiratory symptoms which may be defined differently by different people although there may be difficulty in defining a baseline rate for average bronchodilator use in asthmatics. Many of the studies described are from other countries and the problems of transferability must be appreciated. For PM_{10}, dose-response functions are given in the COMEAP report for asthmatic adults and children for bronchodilator usage, cough and lower respiratory symptoms. A review (Dockery and Pope, 1994) also gives summary dose-response functions for the latter two endpoints. For ozone effects on cough, the COMEAP report noted that the panel study of children in the Six Cities study (Schwartz *et al*, 1994), in view of its size and quality and relevant range of ozone exposure, would be suitable to give an estimate of possible effects. Sulphur dioxide does have effects but is less thoroughly studied (Lebowitz, 1996). Thus, dose-response functions are available if a sensitivity analysis were considered appropriate given the added uncertainties of selecting the baseline population and the appropriate baseline rates.

References

Anderson, H.R., Spix, C., Medina, S., Schouten, J., Castellsague, J., Rossi, G., Zmirou, D., Touloumi, G., Wojtyniak, B., Pönkä, A., Bachárová, L., Schwartz, J. and Katsouyanni, K. (1997) Air pollution and daily admissions for chronic obstructive pulmonary disease in 6 European cities: results from the APHEA project. *Eur. Respir. J.* **10**: 1064-1071.

Burnett, R.T., Dales, R.E., Brook, J.R., Raizenne, M.E. and Krewski, D. (1997) Association between ambient carbon monoxide levels and hospitalizations for congestive heart failure in the elderly in 10 Canadian cities. *Epidemiology* **8**: 162-167.

Burnett, R.T., Dales, R.E., Krewski, D., Vincent, R., Dann, T. and Brook, J. (1995) Associations between ambient particulate sulfate and admissions to Ontario hospitals for cardiac and respiratory diseases. *Am. J. Epidemiol.* **142**: 15-22.

Cortese, A.D. and Spengler, J.D. (1976) Ability of fixed monitoring stations to represent personal carbon monoxide exposure. *J. Air Pollut. Control Assoc.* **26**: 1144-1150.

Department of Health. (1998) Committee on the Medical Effects of Air Pollutants. *Quantification of the Effects of Air Pollution on Health in the United Kingdom.* London: The Stationery Office.

Dockery, D.W. and Pope, C.A. (1994) Acute respiratory effects of particulate air pollution. *Annu. Rev. Public Health* **15**: 107-132.

Katsouyanni, K., Touloumi, G., Spix, C., Schwartz, J., Balducci, F., Medina, S., Rossi, G., Wojtyniak, B., Sunyer, J., Bachárová, L., Schouten, J.P., Pönkä, A. and Anderson, H.R. (1997) Short term effects of ambient sulphur dioxide and particulate matter on mortality in 12 European cities: results from time series data from the APHEA project. *BMJ* **314**: 1658-1663.

Lebowitz M.D. (1996) Epidemiological studies of the respiratory effects of air pollution. *Eur. Respir. J.* **9**:1029-1054.

Ministry of Health (1954). *Mortality and Morbidity During the London Fog of December 1952. Reports on Public Health and Medical Subjects No. 95.* London: HMSO.

Morris, R.D., Naumova, E.N. and Munasinghe, R.L. (1995) Ambient air pollution and hospitalization for congestive heart failure among elderly people in seven large US cities. *Am. J. Public Health* **85**: 1361-1365.

Poloniecki, J.D., Atkinson, R.W., Ponce de Leon, A. and Anderson, H.R. Daily time series for cardiovascular hospital admissions and previous day's air pollution in London, UK. *Occup. Environ. Med.* **54**: 535-540.

Schwartz, J. (1997) Air pollution and hospital admissions for cardiovascular disease in Tuscon. *Epidemiology* **8**: 371-377.

Schwartz, J. and Dockery, D.W. (1992) Increased mortality in Philadelphia associated with daily air pollution concentrations. *Am. Rev. Respir. Dis.* **145**: 600-604.

Schwartz, J., Dockery, D.W., Neas, L.M., Wypij, D., Ware, J.H., Spengler, J.D., Koutrakis, P., Speizer, F.E. and Ferris, B.G. (1994) Acute effects of summer air pollution on respiratory symptoms reporting in children. *Am. J. Respir. Crit. Care Med.* **150(5 Part 1)**: 1234-1242.

Schwartz, J. and Morris, R. (1995) Air pollution and hospital admissions for cardiovascular disease in Detroit, Michigan. *Am. J. Epidemiol.* **142**: 23-35.

Sunyer, J., Spix, C., Quénel, P., Ponce de Leon, A., Pönkä, A., Barumandzadeh, T., Touloumi, G., Bachárová, L., Wojtyniak, B., Vonk, J., Bisanti, L., Schwartz, J. and Katsouyanni, K. (1997) Urban air pollution and emergency admissions for asthma in four European cities: the APHEA project. *Thorax* **52**: 760-765.

Touloumi, G., Samoli, E. and Katsouyanni, K. (1996) Daily mortality and 'winter type' air pollution in Athens, Greece - a time series analysis within the APHEA project. *J. Epidemiol. Community Health* **50(Suppl 1)**: S47-S51.

United States Environmental Protection Agency (1995). *Air Quality Criteria for Particulate Matter. Volume III. EPA/600/AP-95/001c.* Research Triangle Park, NC: US Environmental Protection Agency.

Verhoeff, A.P., Hoek, G., Schwartz, J. and van Wijnen, J.H. (1996) Air pollution and daily mortality in Amsterdam. *Epidemiology* **7**: 225-230.

Wordley, J., Walters, S. and Ayres, J.G. (1997) Short term variations in hospital admissions and mortality and particulate air pollution. *Occup. Environ. Med.* **54**: 108-116.

World Health Organization. *Air Quality Guidelines for Europe 1998.* [In press].

Spix, C., Anderson, H.R., Schwartz, J., Vigotti, M.A., LeTertre, A., Vonk, J.M., Touloumi, G., Balducci, F., Piekarski, T., Bacharova, L., Tobias, A., Pönkä, A. and Katsouyanni, K. (1998) Short-term effects of air pollution on hospital admissions of respiratory diseases in Europe: a quantitative summary of APHEA study results. *Arch. Environ. Health* **53**: 54-64.

Chapter 3

Air Pollution Policy Appraisal - Discussion of Approach

Introduction

3.1 Policy appraisal is the process of identifying, quantifying, weighing up and reporting on the costs and benefits (in their widest sense) of the measures which are proposed to implement a policy - in this case the UK National Air Quality Strategy (Department of the Environment, 1997). For Government Policy, the basic principles are described in the Treasury "Green Book" (HM Treasury, 1997) and in guidance from individual government departments (Department of the Environment, 1991; Department of Health, 1995a).

3.2 The aim of policy appraisal is to ensure that decision-makers are more fully informed about the detailed consequences of a proposed policy and can, thus, make "better" decisions. "Better" decisions may be those that:

- result in more efficient use of scarce resources;

- more adequately reflect public preferences;

- are based on clearer criteria, or more careful use of evidence;

- command wider acceptability amongst those affected;

- are more "transparent" in the sense of being open to scrutiny;

- are most defensible;

- arise from more complete stakeholder participation.

3.3 Appraisal can range from a qualitative discussion of advantages and drawbacks to a full cost-benefit analysis with costs and quantified benefits expressed in monetary terms. A variety of techniques are used to assess the importance of different costs and benefits and to compare them with each other. This Group was not asked to advise on the overall policy appraisal but we have been asked in our terms of reference when advising on how to reflect the importance of the health benefits of reducing air pollution:

- to consider whether monetary valuation of health effects is appropriate in this context;

- to consider the merits of alternative approaches and, if necessary, to recommend further work to help develop the most appropriate approach.

3.4 In this chapter, we present the following:

(i) some background about the perception of risk and how this might affect the importance given to the health effects of air pollution by individuals, experts or society (paragraphs 3.5 - 3.9);

(ii) a discussion of what would constitute an ideal approach (paragraphs 3.10 - 3.12);

(iii) a comparison of different approaches to policy appraisal and how the benefits would be expressed in each case. The approaches (which are not necessarily mutually exclusive) include

- cost-effectiveness analysis (paragraphs 3.16 - 3.19)

- cost-benefit analysis (paragraphs 3.22 - 3.42)

- multicriteria analysis (paragraph 3.43 - 3.48);

(iv) a conclusion on which approach or combination of approaches will be explored further in later chapters (paragraphs 3.49 - 3.55).

Who judges importance of risk - experts, individuals or society?

3.5 A fundamental question in judging the relative importance of benefits concerns whose views are being sought. The general public is affected by decisions on controlling the risks they face and might reasonably be considered to have a right to be involved in those decisions. This immediately raises the issue of whether, and how far, the general public's perceptions of risk *should* be taken into account in these decisions (Pidgeon *et al*, 1992; Okrent and Pidgeon, 1998). There is concern that non-experts may not have the knowledge or resources to evaluate accurately what will harm them, and that there is a potential for resources being diverted from activities which harm most to those that scare most. On the other hand, the public's perceptions do have real consequences when policies are implemented (Kasperson, 1992), as was seen in the BSE controversy, and hence to ignore them entirely might also be damaging. In addition, it is now generally accepted that experts are not totally impartial but work within their own cultural framework (Fischhoff, 1989). Thus, some differences in risk perception between experts and the public are due to differences in attitude rather than misunderstanding of facts. For example, the public may regard a risk that cannot be observed more seriously than one which can, while accepting that quantitative estimates of the risks are the same.

QUALITATIVE FACTORS AFFECTING ATTITUDES TO RISK	
CHARACTERISTIC	DIRECTION OF INFLUENCE
Personal Control	increases risk tolerance
Institutional Control	depends upon confidence (trust) in institutional performance
Catastrophic Potential	decreases tolerance
Voluntariness	increases tolerance
Familiarity	increases tolerance
Dread	decreases tolerance
Inequitable Distribution of Risks and Benefits	depends on individual utility, strong social incentive for rejecting risks
Artificiality of Risk Source	amplifies attention to risk, often decreases risk tolerance
Blame for Cause of Damage	increases quest for social and political response

Source: Renn, 1998

3.6 Social science research has investigated some of the factors which affect people's different attitudes to risk (Pidgeon *et al*, 1992; Pidgeon and Beattie, 1997). We describe these here as they may give some insight into whether the public is likely to attach particular importance to risks from air pollution compared with other risks of similar magnitude. The findings show that the lay concept of risk means more to individuals and groups in society than just expected fatalities or other direct measures of harm. Qualitative characteristics of the hazards (and of the institutional context within which a hazard arises) are also important such as their controllability, voluntariness, dread, catastrophic potential etc., (Slovic *et al*, 1980; Renn, 1998) (see box above). A common example is that people's attitude towards and tolerance of risk depends on whether or not they themselves feel they are in control of it: experience may suggest that the risks associated with two methods of transport one private and one public may be very similar, but people may be far more willing to accept private risks, which they believe they can control and which they recognise are their responsibility, than they are willing to be exposed to the same level of risk[1] from the actions of providers of public transport services. Attitudes to risk are also related to wider social beliefs and contexts (Johnson and Covello, 1987); such as beliefs about equity of distribution of risks and benefits; beliefs about the moral justification (or not) of the risk generating process; blaming others (for risks imposed involuntarily and where consequences could have been foreseen); and the degree of trust placed on some institutions and not others.

3.7 Although there is now more understanding of how these factors influence people's attitudes in general terms, it is less clear whether people are willing or able to "translate" these general attitudes/ethical viewpoints into measurable, stable, consistent, quantitative preferences for particular reductions in risks (Beattie *et al*, 1998; Fischhoff, 1991), such as those associated with air pollution. In part, this is because judgements about such risks involve intrinsically complex and uncertain scientific questions, as well as trade-offs between very different types of values (e.g., between the benefits from mobility and those of human health), but these are also issues which people will generally not have thought about at length or in depth. This is obviously important because approaches to policy appraisal based on assessing public preferences rely on the consistency of these preferences when comparing costs and benefits or the importance of one policy rather than another.

3.8 Several studies have shown that the term "air pollution" elicits significant levels of concern (Health Education Authority, 1997; Kendall, 1997) but it is unclear what is driving these concerns. Some of the concern might be modified by better understanding of the actual risks. For example, it is commonly believed that air pollution causes asthma (rather than just exacerbating symptoms) although this is not supported by the evidence (Department of Health, 1995b). Although it is easy to interpret such findings as demonstrating that the general public are in some sense "irrational" or ill-informed in their thinking about risks, their views are not necessarily unreasonable given the evidence people have at their immediate disposal. Hence, if their children's asthma is indeed triggered disproportionately on days when there are high air pollution levels the distinction between "cause" and "exacerbation" may seem trivial from a practical perspective. Other aspects of concern may be influenced by the characteristics of air pollution, for example because it is an involuntary risk which is not easily observed and it is not easily controlled. The context of air pollution is actually quite complex as there are several different air pollutants, generated through a range of activities meeting different human needs (warmth, mobility etc.), controlled in a variety of ways, and having a variety of effects. All of these aspects may impact on preferences for reducing effects of air pollution in different ways for different groups.

3.9 So far, we have compared the views of individual members of the public with the views of experts. In addition, the aggregate self-interested views of individual members of the public may not match the aggregate views of how others should be treated (the views of society). For example, individuals might not attach much personal importance to allocating resources to preventing problems which will not affect them during their lifetime but society as a whole might regard the principle of protecting future generations as important. Another issue concerns particular sub-groups of the population. A single aggregate measure of individual views may not reflect the difference between a level of importance held fairly evenly across the population and a level of importance derived from widely divergent views. This difference might be important in defining the consequences of policy options. In addition, some groups may be more at risk than others. Should the views of these groups be given more weight than those of others who may contribute resources to control of the risks but do not know what the risks are like or do not need to face the possibility that they might be personally affected? Of course, many of the issues above raise very difficult ethical and policy choices, that are unlikely to be resolved either simply, or to the complete satisfaction of everybody in society.

---

[1] In addition to preferring voluntary risks *per se*, people may also underestimate seemingly controllable risks because they think they can take precautions (e.g., by driving more carefully than average).

Specific approaches to policy appraisal

3.10 We consider that it is difficult to identify unambiguously a "best way" to present the health benefits in isolation since the benefits need to be expressed in a manner that is compatible with the overall approach to policy appraisal. Therefore, in this chapter we describe the main approaches to general policy appraisal and comment on their advantages and disadvantages. Based on this, we then consider which methods of assessing and presenting health benefits merit further investigation in the air pollution context in later chapters.

3.11 We considered what would constitute our ideal approach to the assessment of benefits. The approach would need to be able to consider different health outcomes on a similar scale, for example using differences in survival (life-years) and/or quality of life/morbidity. The approach would also need to lead readily to a comparison of the costs and benefits of alternative air pollution policy measures. In the case of air quality control, decisions are, of course, mainly being taken by government rather than by individuals or companies acting through the marketplace. We are, therefore, concerned with a process of social decision-making and our approach would need to take account of this. Social decisions may be based on the sum of individual views or Governments may decide to override the individual wishes of citizens either in the interests of the collective good or because individuals are thought to be short-sighted or ill-informed. There may also be issues of fairness to take into account, for example policies may be designed to ensure that polluters pay for the control of pollution. There is, thus, a range of information which the government needs in order to arrive at a decision, including information derived from private perceptions of the value to be attached to any change, from the views of society and from scientific and other experts. The approach would also need to allow comparison with other related policy areas. These points are borne in mind in the discussion of the specific approaches.

3.12 Even though this report covers only health effects and not other benefits of reductions in air pollution, there are still too many different effects on health to allow an easy comparison between them. Common sense suggests that more priority should perhaps be given to reducing deaths brought forward than to reducing hospital admissions and, in turn, hospital admissions might be regarded as more important than minor respiratory symptoms. However, it becomes more difficult to judge <u>how</u> much more important one health outcome is than another and how much more important the change in health outcome is than some other outcome of a particular policy (e.g., additional costs for industry). The approaches described below are aimed at assisting this judgement by eliciting and structuring the views people have about the relative importance of different outcomes. Later chapters consider whether data are actually available to allow the use of particular approaches in the air pollution policy appraisal context.

Standard setting and risk criteria

3.13 Standards[2] represent levels (in terms of risk or concentrations of the suspected causal agent) below which the risk is deemed acceptable and above which the risk is deemed unacceptable. Such standards are usually set on the basis of the health evidence alone. Compliance with the standard is then pursued. This appears to avoid trading-off risk against cost. However, the very fact that compliance with the standard is pursued implicitly assumes that this is worthwhile i.e., that the benefits exceed the costs. Because this trade-off is not explicit it can lead to enormous variations in the implied valuations for protecting human health and to an inefficient use of resources. Some standards may be very expensive to meet and may be protecting people against fairly minor effects. Resources would then be better spent pursuing more "efficient" standards. Examples of the extremely wide variation in the resources applied as a result of the use of standards designed to save lives have been illustrated by a number of reviews (e.g., Tengs *et al*, 1995).

---

[2] The term "standard" is not used consistently. It can also be used to describe a level selected <u>after</u> consideration of costs and benefits. We are not referring to this here.

3.14 The air quality standards in the UK National Air Quality Strategy were set purely on health grounds. However, it was always the intention that consideration of costs and benefits should follow as a second stage (Department of the Environment, 1997). This group is contributing to that process.

3.15 In some policy areas "risk criteria" are used in the place of standards, for example, in the context of residents living close to hazardous facilities (Health and Safety Executive, 1989). This simplifies the risk assessment process by defining three levels of risk. High "unacceptable" risks are regarded as requiring immediate action without detailed consideration of costs and benefits, while low "negligible" risks are regarded as trivial and requiring no action. Intermediate risks are subject to consideration of costs and benefits in the sense that levels of exposure should be "as low as reasonably practicable" (ALARP). As with standards, the selection of the thresholds for unacceptable and negligible risk involves an implicit trade-off of risks against costs.

Cost-effectiveness

3.16 Cost-effectiveness analysis consists of the comparison of alternative ways of producing the same or similar outputs, which are not necessarily given a monetary value. It can be used where a budget is fixed and options are compared in order to maximise the amount of benefit gained from that budget. Alternatively, the costs of different options for producing a fixed level of benefit could be compared. For example, the costs per life saved associated with different policies could be considered. However, it should be noted that cost-effectiveness gives no information on whether any of the options being compared should be pursued. (The options might all have more disadvantages than advantages.) Nor does it suggest whether the overall budget allocated to that particular activity is appropriate. Again, therefore, the need to balance risks and costs has not been avoided. Rather, an implicit trade-off has been made when the fixed budget is allocated or the target standard set.

3.17 Cost-effectiveness analysis is widely used in the National Health Service (NHS) to get the best out of the NHS budget. This requires comparison of cost-effectiveness across a wide variety of treatments for different health outcomes. To aid this comparison the effects on health outcomes are expressed in some standardised measure of mortality and/or morbidity improvement. One variant of this which is becoming more widely used is to describe the health effects in terms of changes in "quality-adjusted life years" or QALYs (Department of Health, 1995a). This can be used to reflect changes in morbidity and mortality. Quality-adjusted life years are a composite measure where the estimated life years gained (or lost) are weighted according to expected quality of life in those years. Years of good health are regarded as more desirable than years of poor health. The degree of poor health can be measured using a number of different scales (Kind, 1989). For conversion into quality-adjusted life years, the most appropriate scales are those where illnesses are defined in terms of quality of life states based on dimensions such as pain, anxiety and disability and these quality of life states have been given a relative value by a representative sample of the general population using 1 to represent full health and 0 to represent death. Although it is only a recent development, the EQ5D (EuroQol Group, 1990; Brooks, 1996; Williams, 1997a) (Annex 3A) is now widely regarded in the UK as the most appropriate scale to use. Another scale, the Quality of Well Being (QWB) scale (Patrick *et al*, 1973; Kaplan *et al*, 1993) (Annex 3A), is more often used in the US. The score is then used to weight the life years. For example, 10 years in a quality of life state rated at 0.5 would equal 5 quality-adjusted life years. The cost of a particular treatment could then be compared with the quality-adjusted life years gained to derive a "cost per quality-adjusted life year"

3.18 Generic measures of quality of life may be too insensitive to detect clinically significant changes in particular diseases. Disease specific measures may provide a more sensitive measure of quality of life related to particular diseases (Bowling, 1995). (The St George's Respiratory Questionnaire (Jones et al, 1992; Jones, 1993) is one example we mention later in the report[3].) However, the scales do not necessarily cover the same range (from full health to death) as for the generic measures so it is difficult to convert between the two. In addition, they cannot be used in situations where more than one disease is affected. Unfortunately, very little work has been done to compare the disease-specific measures and more general measures within the same studies (Harper et al, 1997).

3.19 The quality of life states are scored without reference to the disease responsible for that state. This aids comparison and could be regarded as an advantage on equity grounds since diseases causing similar degrees of suffering are treated equally. On the other hand, the scoring will not take account of the differential importance attached by individuals to avoiding a given loss in quality of life from different diseases. People may, for example, have a particular dread of cancer. The quality of life states are also presented out of the context of the age of those affected. As currently used, a quality-adjusted life year gained for somebody at age 20 is treated the same as a quality-adjusted life year gained for somebody at age 70, and 1 person gaining 10 quality-adjusted life years is regarded as the same as 10 people gaining 1 quality-adjusted life year each. (The invariance with age has been challenged (Williams, 1997b) and methods of differential valuation are being developed (Dolan and Green, 1998).) In addition, a quality-adjusted life year gained by someone in a very severely impaired health state is treated the same as a quality-adjusted life year gained by someone in only mildly impaired health. Finally, the empirical estimation of quality of life scores makes somewhat restrictive assumptions about people's underlying preferences (assuming a zero discount rate, for example) (Broome, 1993; Dolan and Jones-Lee, 1997) and about the effect of duration on valuation.

Risk-risk analysis

3.20 In risk-risk analysis, people may be asked how they would trade-off one risk (e.g., death in a road accident) against another (e.g., risk of developing chronic bronchitis) (Viscusi et al, 1991). It can be used as part of a survey method to assist in eliciting respondent's relative preferences for reducing different types of risks and can derive monetary values for reduction of one type of risk by comparison with another risk for which the monetary valuation has already been established. It can also be used as part of standard setting. For example, the US EPA may consider risk-risk trade-offs when setting maximum concentration limits (MCLs) for contaminants under the Safe Drinking Water Act (US EPA, 1996). The MCLs may be changed to minimise overall risks by balancing the risks from the contaminant with the possible side-effects of the treatments used to reduce levels of the contaminant.

3.21 More ambitious applications have involved the use of risk-risk analysis to compare the benefits of health and safety legislation against the risks that increased "costs to industry" place on economic prosperity and hence on health (Cooper and Nye, 1995). Such analyses have argued that requiring industry to invest in pollution control retards the nation's economic prosperity and, if one accepts that there is a positive correlation between GDP and health, that this loss of prosperity will in turn increase risks to health. The policy aim should, therefore, be to balance the *direct health benefits* of environmental measures (lowering of health risks through pollution reduction) with the *indirect health costs* (increasing health risks due to foregoing economic prosperity) to arrive at the point where the overall risks to health are minimised. Though this type of analysis has attractions, a number of issues arise in its practical application. For example, the relationship between GDP and the health of the population is open to dispute (some diseases might increase with increased prosperity), as is that between "costs to industry" and GDP. As a result, this use of risk-risk analysis will not be considered further here.

---

[3] One study (Harper et al, 1997) which compared several measures suggested that the completion rate of some items on the St George's Respiratory Questionnaire was lower than for other questionnaires.

Cost-benefit analysis and monetary valuation

(i) *The rationale for willingness to pay (WTP)-based monetary values of safety* (Jones-Lee and Loomes, 1998)

3.22 Two hard facts confront those who have to make decisions about the appropriate level of provision of public safety. First, safety is usually not costless; and second, society has limited resources. Consequently, a responsible decision about any proposed public safety improvement will require a judgement as to whether the resulting reduction in risk is large enough to justify incurring the cost of implementing the improvement. Put another way: is the reduction in risk resulting from the allocation of scarce resources to a particular safety project worth more than whatever other good things could be provided if those resources were diverted elsewhere?

3.23 How easily can such a judgement be made? Intuitively, most of us might agree that a safety improvement costing just a few thousand pounds and which could be expected to prevent a number of premature deaths, would be well worth it. Equally, most people would feel that if it would cost several millions of pounds to prevent, at best, only a few minor injuries, then there would probably be many better ways to spend the money. However, in the less extreme cases, which are more typical, the decision may not be quite so straightforward.

3.24 Clearly, if it were possible to obtain an acceptable measure of the *monetary value of safety*, then this would go a long way towards resolving the difficulty. Given such a measure, safety improvement benefits could be weighed explicitly against other costs and benefits - such as capital costs and time savings - in reaching a decision for or against any particular safety project. Indeed, without an explicit measure of the monetary value of safety, serious inconsistencies are likely to emerge in the decision-making process. But how do we arrive at such monetary values of safety?

3.25 The key to this question is that members of the public not only stand to benefit from improved public safety, but also ultimately *pay* for it (either directly through, say, fares on public transport, or indirectly through taxation). If social decisions are to take account of the public's preferences, then this suggests that values of safety should reflect the rate at which members of the public are willing to trade-off safety against other desirable things that might be purchased. In short, there is a very persuasive case for basing values of safety for use in public sector project appraisal on people's *collective willingness to pay for it*.

3.26 Under this WTP approach to the valuation of safety we should, therefore, ideally like to discover how much members of the public would be willing to pay for improvements in their own safety[4] (Where deaths might be involved, the improvements in safety are expressed as small reductions in people's *risk* of death - people are not asked to value individual lives - see paragraph 3.31). The total sum elicited would then be a clear reflection of what the safety improvement was worth to people in the affected group, relative to alternative ways in which they could have spent their limited incomes.

3.27 Most people readily accept that the best way to observe the monetary value which people put on a visit to the cinema or a new outfit is to observe how much they are willing to pay for these items of consumption. However, some people may have the view that lives cannot be treated as tradeable in the same way as goods bought in shops (although this is not what is being asked). Some people may also be sceptical of the likelihood that people will be able to express reasoned and consistent choices concerning the value they put on options which increase or reduce the statistical risk of death from various causes. Nonetheless, in their daily lives, people

---

[4] It should be noted that throughout the report we focus exclusively upon people's valuation of their own safety and make no attempt to capture their altruistic concern, and possible WTP for other people's safety. The reason for this is that recent theoretical work has shown that it is far from clear that WTP-based values of safety intended for use in public sector decision making should take any account at all of people's WTP for others' safety - see, for example, Jones-Lee (1992). Roughly speaking, the reason for this is that for certain quite plausible types of altruistic concern in which people respect others' preferences (commonly referred to in the literature as "pure" altruism), inclusion of WTP for other people's safety would involve a form of "double-counting". It should also be noted that so far attempts to determine the nature of altruistic concern empirically (i.e., to establish whether or not it is predominantly "pure" or by contrast, "paternalistic") have proved to be largely unsuccessful.

do accept the risk that some of their actions could lead to death because they want other things such as the income to enjoy material things or to travel to see new places. Many occupations involve risks of injury or even death and most forms of travel entail some, albeit generally remote, risk of harm. In general, most people tolerate some risks in order to attain things they enjoy or need. All this suggests people do trade-off risks against other things that are of value to them. The more difficult issue of whether they are able to express this "value" in monetary terms is discussed below in paragraphs 3.37/3.38.

3.28 The values of safety that emerge from this approach will tend to be lower for groups of people with below average incomes and higher for those who are better off, and people may object to the implications of this. It is essentially for this reason that most advocates of the WTP approach would recommend the application of *uniform* values of safety, reflecting the aggregate WTP of a *representative sample* of the population as a whole.

3.29 Those who benefit from safety improvements and those who pay for them may not necessarily be the same group. However, the justification for using WTP values does not depend on those who benefit being the same people as those who pay. The key issue is the overall size of the benefit - and being able to compare it with the cost. Whether the beneficiaries pay or others pay may be resolved by reference to other criteria - such as fairness. This can be dealt with separately from measuring the size of the benefit.

(ii) *Estimating WTP-based values of safety - general principles*

3.30 Consider a road safety improvement that is expected to reduce each person's risk of premature death during the coming year by an average of 1 in 100,000. This risk reduction would mean, on average, that, in a group of 100,000 people, the number of premature deaths during the coming year would be reduced by one. Now suppose that members of the population concerned are, on average, each willing to pay £10 to effect the safety improvement. This means that for each death prevented, there are 100,000 people willing, between them, to pay £10 x 100,000. In this situation where, statistically, 1 life would be saved in the group of 100,000, the aggregate WTP per death prevented would be £1m. It has become the convention to refer to such a result as the *"value of a statistical life"* (or "VoSL"). The expression is open to misinterpretation - and, indeed, it is not really accurate since it is not about valuing 1 life, but about the aggregate value that a large group of people places on typically very small reductions in the risk faced by each individual member of the group. In this report, we have used the alternative term the *"value of prevention of a statistical fatality"* (or VPF)[5] (This does of course refer to premature fatalities since eventually everyone will die.)

3.31 It should be clearly understood that this approach cannot be regarded, in any sense, as placing a value upon the life of a particular individual. We are not valuing loss of life *ex post*, or assessing the resources which should be spent in saving or prolonging individual lives. We are looking *ex ante* at how people view reductions in the risk of death. Though this approach has been criticised[6] (Broome, 1985; Broome, 1991), it seems appropriate to cases where policy decisions will *prospectively* affect risks that are thinly spread across a wide population. If it is impossible to know, at the time the decision is to be made, who will eventually suffer, and how badly, the decision can only be based on *ex ante* values. *Ex post* valuations are simply unavailable.

3.32 Clearly, in the above example, the average individual WTP, £10, for the average individual risk reduction of 1 in 100,000 is a reflection of the rate at which individuals in the group concerned are willing to trade off wealth for risk, and it is, therefore, upon these individual wealth/risk trade-off rates that empirical work in the safety field tends to focus. Of course, the use of

---

[5] An alternative but equivalent way to calculate the VPF is to appreciate that if a total of N people are at risk and would benefit from the safety improvement, then the overall number of fatalities prevented would be N x 1/100,000. In turn, aggregate WTP would be £10 x N and aggregate WTP per fatality prevented would be £10 x N divided by N/100,000 i.e., it would again be given by £10 x 100,000.

[6] A longstanding critic of this approach is Broome (Broome, 1985; Broome 1991), who has difficulties with various aspects, including the failure to take account of effects on the potential progeny of those whose lives are at risk, and indeed the whole notion that people's preferences indicate what is good for them. However, further discussion of such issues is beyond the scope of this report.

average values can mask considerable individual variation, particularly in terms of attitudes to risk. There will probably be some people in the group who would rather face the extra risk than pay £10 while others, who are more "risk averse", would rather have paid more for a greater reduction. Where policy decisions affect the whole population, however, the use of average values may provide the most appropriate guidance. Empirical work in the safety valuation field, therefore, tends to focus on the aggregate results of individual wealth/risk trade-off rates.

3.33 Broadly speaking, procedures for obtaining empirical estimates of WTP-based values for the prevention of fatalities and non-fatal injuries can be classified into one of three types, namely:

the revealed preference (or implicit value) approach;

the contingent valuation (or questionnaire) approach;

the relative valuation approach.

(iii) *Estimating WTP-based values of safety - the revealed preference approach*

3.34 Essentially, the revealed preference[7] approach involves the identification of situations in which people actually do trade off income or wealth for physical risk, typically - though by no means invariably - in labour markets where riskier jobs can be expected to command clearly identifiable wage premiums. The aim is then to estimate the rate at which individuals are willing to trade off wealth or income against the risk of death or injury on the basis of data concerning the choices that they actually make in the situation concerned. Thus, in the case of labour market studies, these trade-off rates would be estimated from, *inter alia*, observed wage rates and occupational risk levels.

3.35 While the revealed preference approach has the advantage of dealing with actual choices, when applied to labour market data it suffers from the disadvantage that wage rates can be expected to depend on many factors besides occupational risk, so that it is necessary to disentangle those other effects in estimating the relationship between wage rates and job risk. Also, workers may or may not be well-informed as to the risks they run. Furthermore, it is also the case that observed wage-risk relationships may owe as much to the intervention of regulatory bodies, such as the UK Health and Safety Executive, as they do to the interaction of workers' preferences and employers' profit motives. Thus, what one may be picking up in wage-risk studies is the rate at which *regulatory bodies* are willing to allow employers to trade workers' wealth for risk, rather than the marginal rate which is, in fact, acceptable to workers.

(iv) *Estimating WTP-based values of safety - the contingent valuation approach*

3.36 By contrast with the relative complexity of the revealed preference approach, the contingent valuation approach involves asking a representative sample of people more or less directly about their WTP for improved safety (or sometimes their willingness to accept compensation for increased risk). Thus, the contingent valuation approach allows the researcher to address explicitly and unambiguously the required wealth/risk trade-off, at least in principle. In addition, under the contingent valuation approach it is a straightforward matter to establish the way in which people's valuation of safety is affected by factors such as income, age, social class and other demographic characteristics, whereas the revealed preference approach necessarily operates at a far more highly aggregated level, focusing on market equilibrium wealth/risk trade-offs.

---

[7] We noted previously that standard setting and cost-effectiveness involved implicit rather than explicit trade-offs and that this could lead to an inefficient use of resources. In these cases, the implicit trade-offs are probably an accidental result of the other considerations of the regulatory bodies (hence the wide variation). In the revealed preference approach when wage premia for risky jobs are examined, it is more likely that the individuals have considered the wages and the risks of the job and thus, at least in theory, a genuine trade-off is being revealed.

3.37 On the other hand, the contingent valuation approach has some problems and has generated a good deal of controversy. Thus, for example, it relies on answers to questions concerning essentially *hypothetical* choices and there is clearly a danger that respondents may, for one reason or another, systematically misrepresent their true preferences or be subject to unconscious biases. However, there is very little hard evidence that deliberate misrepresentation is pervasive. Of more concern, perhaps, is the point that most people have very little direct experience of the kind of explicit money/risk trade-offs that are typically posed in contingent valuation questions. It may therefore, as noted earlier in paragraph 3.7, be over-optimistic to expect them to be able to give thoroughly considered and stable answers to such questions.

3.38 More specifically, there is evidence that in some cases responses to contingent valuation questions may be unduly influenced by the precise way in which such questions are worded (often referred to as "framing" effects) and may also be *insufficiently* sensitive to the quantity of the good(s) on offer in the question concerned ("embedding" or "scope" effects) (Beattie *et al*, 1998; Cummings *et al*, 1986; Kahneman and Knetsch, 1992; Arrow *et al*, 1993; Bjornstad and Kahn, 1996). There is also concern that the results are not always reproducible in a repeated survey (Royal Commission on Environmental Pollution, 1998). Nonetheless, there are good grounds for believing that the more carefully conducted contingent valuation studies (for example Chilton *et al*, 1998) do give an adequately reliable indication of at least the broad order of magnitude of the value that members of the public place on safety improvement. Thus, provided that it is appreciated that such studies are, by their nature, unlikely ever to be capable of generating very precise point-estimates but are, rather, best viewed as the source of "broad-brush" indicators, then one can be reasonably confident that these studies are adequate for their intended purpose.

(v) *Estimating WTP-based values of safety - the relative valuation approach*

3.39 Finally, what of the relative valuation approach? In contrast to the revealed preference and contingent valuation approaches, this does *not* involve an attempt to estimate wealth/risk trade-off rates directly, but rather seeks to determine the value of preventing one kind of physical harm *relative to* another. Thus, for example, the DETR's current monetary values for the prevention of different severities of non-fatal road injury were obtained by applying such relative valuations to an absolute monetary peg in the form of the DETR's WTP-based monetary value for the prevention of a road fatality.

3.40 In fact, several different procedures have been employed to estimate relative valuations, but to the extent that all of these involve asking a representative sample of the population at risk questions relating to hypothetical choices or trade-offs, they clearly have very much more in common with the contingent valuation approach than with revealed preference.

3.41 More specifically, relative valuation questions typically seek to determine *either* the rate at which respondents are willing to trade off the risk of one kind of physical harm against the risk of another kind [so-called risk-risk or standard gamble questions (Viscusi *et al*, 1991; Jones-Lee *et al*, 1995)] *or* the number of casualties of one type whose prevention would be regarded by respondents as being "equally as good" as the prevention of a given number of casualties of another type [so called "matching" or "person trade-off" questions (Jones-Lee and Loomes, 1995)].

3.42 At least some of the concerns people have about WTP and particularly the "value of prevention of a statistical fatality" are related to misinterpretations of these concepts. We hope that these descriptions of the various methods of eliciting WTP values will assist in a fuller and better understanding of what the approach actually involves.

Multi-criteria analysis

3.43 Establishing WTP represents one approach among many toward investigating how individuals value different things, and aggregating these individual preferences to help illuminate policy choices. An extensive review of such methods, listing the advantages and potential pitfalls of each illustrated in the context of health care provision, is provided by Mullen (1998). Similar approaches can in principle be used to investigate attitudes toward public health issues such as air pollution. Given the remit of this report, WTP is of particular interest as the best-researched way of establishing monetary valuations which might then be used to inform wider benefit-cost analysis. Nevertheless, monetary valuation is not the only way of seeking to clarify the trade-off between the different things that people want.

3.44 In particular, it is important to distinguish two ways of structuring choices. One seeks to measure any cost or benefit on the same single scale (usually, though not necessarily, through monetary valuation). The other approach is to measure different effects on different scales, e.g., monetary cost in terms of money, health risks in terms of expected health outcomes, and to devise a way of choosing between them, for example by prioritising or "weighting" objectives (Watson and Buede, 1987). These are known generically as *multiple criteria* methods. Such methods can be applied formally, with explicit scoring of each of the different, apparently incommensurate, elements. However, the same approach can be used less formally. In reality, a need for a consideration of multiple factors is probably inevitable when developing policies to reduce risks to public health since rather few decisions involve a straight trade-off between health risks and money (and *nothing* else). This suggests that any form of political decision-making can be seen as a form of multi-criteria analysis (just as it can also be seen as an informal form of cost-benefit analysis since it will involve consideration of advantages and disadvantages). In everyday life, as we have already noted, people routinely trade off personal safety against a variety of other *non-monetary* criteria such as enjoyment, or time or convenience. In doing so, they "solve" a multiple criteria decision problem - without going through the process of putting a monetary value on safety, another on enjoyment, time or convenience and then comparing the answers. Various forms of analysis are available to structure this process in more complex areas, two of which are described below.

(i) *Conjoint analysis*

3.45 Conjoint analysis aims to establish the relative importance placed on different criteria when people evaluate particular goods or services. Having been used quite widely in market research, it has more recently been applied to provision of health care - of obvious relevance here in the sense of relating health outcomes and other criteria (Ryan, 1996a). A typical study would first investigate what criteria most people considered significant in judging a particular issue. They are then presented with hypothetical scenarios that are better in some respects and worse in others and asked to rate the scenarios against an overall scale of desirability, rank them from best to worst, or express a preference between different pairs of scenarios. Each method has relative advantages: for example, the third poses respondents with the simplest questions, but more questions are needed in order to get the same information. Whichever is used, the point is that information is gained about respondents' *willingness to trade one desirable attribute for another*. Given a large-scale study, statistical analysis can establish the average "rate of substitution" for any pair of criteria - for example, how much a shorter stay in hospital would compensate for some extra risk of a given side-effect.

3.46 Conjoint analysis methods recognise the multi-faceted nature of choice and seek to find the criteria that actually matter to those faced with a risky situation. However, they have some limitations as compared with WTP surveys. In particular, since the "rates of substitution" found are only averages emerging from statistical analysis, one cannot test their validity by investigating the consistency of individual answers. Some have argued that the two approaches should, therefore, be used in combination. Indeed, if the scenarios include cost as a variable, the results can be analysed to find the *implied* WTP for any of the other attributes, including risk reduction. This has in fact been done recently in the air pollution context (Diener *et al*, 1997) - this study is considered further in Chapter 6. One can, therefore, see whether the two methods yield compatible results as to the implied "value of prevention of a statistical fatality". In at least some cases, they do appear to give similar results (Ryan, 1996b).

(ii)   Multiple criteria decision analysis (MCDA)

3.47   While conjoint analysis principally investigates people's views about multiple criteria decisions, MCDA is primarily conceived of as a means of assisting decision-makers (Watson and Buede, 1987; Belton, 1990). It is not really an alternative to cost-effectiveness or cost-benefit analysis since it can incorporate information derived by these other methods. Decision analysis itself models decisions in which the consequences of alternative options may be uncertain, but the uncertainty can - at least roughly - be expressed in terms of probabilities. It thus fits naturally into discussion of decisions about risk. In its simplest form, decision analysis deals with cases where there is a single objective but the *multiple criteria* version does away with this. Instead, it deals with cases in which there are several objectives, liable to be at least partly competing. As with conjoint analysis, cost can be included as one dimension. Analysis using MCDA would involve:

- establishing a hierarchy of criteria to represent the decision-maker's values;

- defining the alternative options available;

- scoring each alternative to show how it performs against each criterion (where necessary using probabilities, as in "simple" decision analysis);

- giving the various criteria *weights* to represent their relative importance; and

- multiplying scores and weights to calculate the overall weighted score for each alternative.

3.48   Within this overall approach many specific techniques have been developed. Several make use of flexible, interactive software which allows the decision-models to be displayed, analysed and changed very readily. The models can, thus, be used to focus discussion, for example through trying out different weights for the various criteria to explore the effects on the final result. This can also help clarify which are the most critical areas of uncertainty needing further investigation. Whereas conjoint analysis is useful for asking large numbers of people comparatively simple questions, MCDA is more typically used to explore issues in greater depth with fewer people - often decision-makers and their advisers. This may seem less democratic but by making value judgements more explicit, use of MCDA has the potential for making decisions more transparent and accountable (provided decision-makers are willing - or required - to make the model public). Used in this role, multiple-criteria models have the advantage of allowing costs and benefits to different parties to be clearly differentiated (by taking them as separate criteria) rather than obscured within a measure of overall social welfare. Although less common, MCDA can be used to capture the views of a wider public by structuring the proceedings of extended discussion groups with members of the public or representatives of stakeholder groups (Gregory *et al*, 1993; Ives and Footitt, 1996).

Conclusions

3.49   In summary, standard setting and risk criteria, while useful for other purposes, do not examine costs and benefits explicitly. Other approaches do examine costs and benefits directly but not always on a comparable scale. We have been asked to consider the best way to express the importance of the health benefits from reductions in air pollution for use in a cost-benefit assessment. The options for expressing the benefits before comparison with the costs are:

(a) measuring health benefits in non-monetary units :

   (i) ranks or scores as in conjoint analysis or multi-criteria decision analysis;

   (ii) standardised health outcome measures including measures of quality of life and life expectancy;

(b) measuring health benefits in monetary units such as estimates of WTP for reductions in health risks.

A summary of the advantages and disadvantages of these options is given in Table 3.1.

Table 3.1 **Comparison of possible approaches to policy appraisal**

Criterion	Cost-effectiveness	Cost-benefit analysis - monetary valuation	Multi-criteria analysis (can be combined with other two)
Different health outcomes on similar scale	Standardised health outcome measure e.g., quality/duration of life.	Monetary scale	Abstract scale of scores
Ready comparison of costs and benefits so can determine if action is justified by benefits exceeding the costs.	Can compare quality and duration of life per unit cost but cannot determine if benefits exceed the costs.	Costs and benefits both in monetary units so straightforward to determine if the benefits exceed the costs.	Could express costs and benefits as scores but cannot determine if the benefits exceed the costs in resource terms.
Compatible with techniques usually used in relevant policy area (health or environment).	Measurement of health gain in terms of quality of life/ life expectancy increasingly used in the health service. Other forms of cost-effectiveness used in environmental policies.	Not generally used in the health service but standard technique for assessment of many government policies including environmental policies with an impact on public health.	Currently not widely used in either the health or the environment policy context.
Takes account of individuals' views	Quality of life states scored by individuals but other aspects e.g., dread of particular diseases not taken into account.	Based on individual views but resultant average values applied uniformly to all income groups.	May use views of experts rather than lay people
Takes account of views of society (e.g., equity), (sometimes overrides individual views)	Loss of quality of life treated the same regardless of age or type of disease	Only by applying common value to all income groups.	Could include e.g., equity as a criterion
Approach reasonably well developed	Yes.	Yes.	Conjoint analysis recent technique.
Others	Some problems with underlying theoretical assumptions.	Should ensure responses not unduly influenced by differences in wording of questions. Can be perceived as "caring only about money"	Can use with CBA or CEA. If only use experts may not reflect wider views.

3.50 In discussing an ideal approach, we noted the importance of an easy comparison between costs and benefits. This criterion is not fully met by scoring systems or health outcome measures although they are useful for cost-effectiveness analysis. We also note that some scoring systems, particularly conjoint analysis, have been only relatively recently developed and they have not been used extensively in the public safety area. For these reasons, we do not discuss scoring systems further in later chapters to any extent although we give some discussion of how Multiple Criteria Decision Analysis might be used in Annex 3B and discuss one conjoint analysis study in later chapters. Although such studies will not show whether the benefits exceed the costs or vice versa, they can be useful in further clarifying the trade-offs to be made in the area of air pollution.

3.51 If the choice is to use resources in reducing pollution and thus reducing effects on health rather than spending equivalent resources on improving health in other ways, then it seems clear that the best way in which the cost-effectiveness of different options can be assessed is via the assessment of the impact on quality of life and life expectancy. This would ensure consistency with consideration of other policy areas designed to improve health. We are a group constituted by the Department of Health and we expect that an important audience for our report will be health professionals. Given that people in the health service are increasingly familiar with quality-adjusted life years and coming to understand the concept with its uses and limitations, we believe that, despite the central focus on the need for a full cost-benefit analysis, our work should also provide pointers to the implications for quality-adjusted life years. We have, therefore, introduced this second thread of analysis throughout the detailed chapters of our report.

3.52 We have also examined whether we could combine the WTP approach (which allows direct comparison with the costs of pollution control) and the quality-adjusted life year approach. For example, the average loss of life expectancy for a death in a road accident can be estimated and expressed in quality-adjusted life year terms (Department of Health, 1995a). The WTP to reduce the risk of a death in a road accident can then be expressed as the WTP to reduce the risk of loss of a certain number of quality-adjusted life years. This would lead to a WTP for a quality-adjusted life year. However, it should be noted here, that WTP and quality-adjusted life years were developed separately and rely on different assumptions. There are also theoretical reasons why they may be difficult to combine. This can lead to some inconsistencies although not all of the assumptions are essential to each technique. Willingness to pay for a quality-adjusted life year is discussed further in Chapter 4.

3.53 We have noted already that the WTP approach is the main approach which allows determination of whether the benefits exceed the costs or vice versa. Of course, some of the benefits of reducing air pollution such as reduced healthcare costs and reduced loss of productivity (see Chapter 5) can already be compared with the costs of pollution control as they are in monetary terms. However, comparing these alone, while useful, would omit the grief and suffering caused by the health effects (i.e., the disbenefit of the decline in health itself). We believe it is important that such distress should be taken into account in a full cost-benefit analysis and that this is difficult to do if it is on a different scale. Therefore, we examine the monetary valuation approach further in later chapters. We will continue to bear in mind some of the issues raised in this chapter - the concern over whether individuals actually hold stable quantitative preferences (paragraph 3.7), the need to understand what drives people's perceptions of the importance of different risks (paragraph 3.6) and whether some of the aspects determining individuals' views (e.g. their income) need to be overridden in the interest of society's principles (paragraph 3.9). In particular, we note that, due to the different perception of risks in different contexts (paragraph 3.6), it is likely that WTP values specific to the air pollution context will be needed.

3.54 Whatever approach is used, there is a need for more systematic research on: (1) the levels of knowledge and understanding people currently hold on the causes and effects of air pollution; (2) what extra knowledge people might need to hold in order to make informed decisions about their personal health protection with respect to air pollution; and (3), in relation to the central question of valuation, what processes of deliberation (Stern and Fineberg, 1996) and information provision might help people to construct stable and consistent preferences in relation to risk reduction in this complex scientific domain. Here more qualitative and intensive approaches to value elicitation hold considerable potential, in some form of hybrid method drawing upon techniques from structured group discussions, and the more formal decision analytic methods discussed above in 3.44. Such approaches are useful for eliciting what people actually know and find important about air pollution risks in their own terms (rather than in terms pre-defined by scientists or in a structured value elicitation questionnaire), and have scope for provision of relevant information to help people form considered preferences, as well as inclusion of consistency checks on the stability of such preferences.

3.55 This Chapter has concentrated on the general principles of the various approaches each of which has advantages and disadvantages. Given the problems associated with each of the methods, the most fruitful approach in the longer term may well be to use the different methods alongside each other, and to cross-check the results obtained. For the present, WTP surveys provide the most well-developed source of empirical data about the *stated* trade-offs between health risk reductions and cost on the part of the general public. We have concluded that the WTP approach and the quality-adjusted life year approach merit further consideration of whether they can be applied in practice in the context of air pollution policy appraisal. This is done in the following chapters.

References

Arrow, K., Solow, R., Portney, P., Leamer, E., Radner, R. and Schuman, H. (1993) Report of the NOAA Panel on Contingent Valuation. *Fed. Reg.* **58**: 4602-4614.

Beattie, J., Chilton, S., Cookson, R., Covey, J., Hopkins, L., Jones-Lee, M., Loomes, G., Pidgeon, N.F., Robinson, A. and Spencer, A. (1998) *Valuing Health and Safety Controls: A Literature Review*. London: HSE Books.

Belton, V. (1990) Multiple criteria decision analysis - practically the only way to choose. In: *Operational Research Society Conference Tutorial Papers*. (Hendy, L.C. and Eglese, R. eds). Birmingham: Operational Research Society.

Bjornstad, D.J. and Kahn, J.R. (eds). (1996) *The Contingent Valuation of Environmental Resources: Methodological Issues and Research Needs*. Cheltenham, UK; Brookfield, US: Edward Elgar.

Bowling, A. (1995). *Measuring Disease: A Review of Disease-Specific Quality of Life Measurement Scales*. Buckingham, Philadelphia: Open University Press.

Brooks, R. and EuroQol Group (1998). EuroQol: the current state of play. *Health Economics* **7**: 207-312.

Broome, J. (1991) *Weighing Goods*. Oxford: Blackwell.

Broome, J. (1993) Qalys. *J. Public Econom.* **50**: 149-167.

Broome, J. (1985) The economic value of life. *Economica* **52**: 281-294.

Chilton, S., Covey, J., Hopkins, L., Jones-Lee, M.W., Loomes, G., Pidgeon, N. and Spencer, A. (1998) New research results on the valuation of preventing fatal road accident casualties. In: *Road Accidents Great Britain 1997*. London: The Stationery Office.

Cooper, A. and Nye, R. (1995) *What price a life. The use and reform of risk assessment in Government*. Hard Data Paper No 5. London: Social Market Foundation.

Cummings, R.G., Brookshire, D.S. and Schulze, W.D. (eds). (1986) *Valuing Environmental Goods*. Totowa, NJ: Rowman and Allanheld.

Department of the Environment (1997). *The United Kingdom National Air Quality Strategy*. London: The Stationery Office.

Department of the Environment (1991). *Policy Appraisal and the Environment: A Guide for Government Departments*. London: HMSO.

Department of Health. (1995a) *Policy Appraisal and Health. A Guide from the Department of Health*. London: Department of Health.

Department of Health. Committee on the Medical Effects of Air Pollutants (1995b). *Asthma and Outdoor Air Pollution*. London: HMSO.

Diener, A.A., Muller, R.A. and Robb, A.L. (1997) *Willingness-to-Pay for Improved Air Quality in Hamilton-Wentworth: A Choice Experiment*. Working Paper No 97-08. Hamilton, Ontario, Canada: Department of Economics, McMaster University.

Dolan, P. and Jones-Lee, M.W. (1997) The time trade-off: a note on the effect of lifetime reallocation of consumption and discounting. *J. Health Econom.* **16**: 731-739.

Dolan, P. and Green C. (1998) Using the person trade-off approach to examine differences between individual and social values. *Health Economics* **7**: 307-312.

EuroQol Group (1990). EuroQol: A New Facility for the Measurement of Health Related Quality of Life. *Health Policy* **16**: 199-208.

Fischhoff, B. (1989) Risk: a guide to controversy. In: *Improving Risk Communication*. Appendix C. National Research Council Committee. pp.211-319. Washington, DC: National Academy Press.

Fischhoff, B. (1991) Value elicitation: is there anything in there? *Am. Psychol.* **46**: 835-847.

Gregory, R., Lichtenstein, S. and Slovic, P. (1993) Valuing environmental resources: a constructive approach. *J. Risk Uncertainty* **7**: 177-197.

Harper, R., Brazier, J.E., Waterhouse, J.C., Walters, S.J., Jones, N.M.B. and Howard, P. (1997) Comparison of outcome measures for patients with chronic obstructive pulmonary disease (COPD) in an outpatient setting. *Thorax* **52**: 879-887.

Health Education Authority (1997). *Air pollution. What People Think About Air Pollution, Their Health in General, and Asthma in Particular*. London: Health Education Authority.

Health and Safety Executive (1989). *Risk Criteria for Land-use Planning in the Vicinity of Major Industrial Hazards*. London: HMSO.

HM Treasury (1997). *Appraisal and Evaluation in Central Government: "The Green Book"*. London: The Stationery Office.

Ives, D.P. and Footitt, A.J. (1996) *Risk Ranking. Final report submitted to HSE Research Strategy Unit RSU 3444/R71.011*. University of East Anglia: Centre for Environmental and Risk Management.

Johnson, B.B. and Covello, V.T. (eds). (1987) *The Social and Cultural Construction of Risk*. Dordrecht, The Netherlands: Reidel.

Jones, P.W. (1993) Measurement of health-related quality of life in asthma and chronic obstructive airways disease. In: *Quality of Life Assessment: Key Issues in the 1990s*. (Walker, S.R. and Rosser, R.M. eds). Dordrecht/Boston/London: Academic Publishers.

Jones, P.W., Quirk, F.H., Baveystock, C.M. and Littlejohns, P. (1992) A self-complete measure of health status for chronic airflow limitation - the St George's Respiratory Questionnaire. *Am. Rev. Respir. Dis.* **145**: 1321-1327.

Jones-Lee, M.W. (1992) Paternalistic altruism and the value of statistical life. *Economic J.* **102**: 80-90

Jones-Lee, M.W. and Loomes, G. (1998) *Some Notes on the Value of Preventing a Road Fatality*. University of Newcastle upon Tyne: Mimeo.

Jones-Lee, M.W. and Loomes, G. (1995) Scale and context effects in the valuation of transport safety. *J. Risk Uncertainty* **11**: 183-203.

Jones-Lee, M.W., Loomes, G. and Philips, P.R. (1995) Valuing the prevention of non-fatal road injuries: contingent valuation vs standard gambles. *Oxford Economic Papers* **47**: 676-695.

Kahneman, D. and Knetsch, J.L. (1992) Valuing public goods: the purchase of moral satisfaction. *J. Environ. Econom. Manage.* **22**: 57-70.

Kaplan, R.M., Anderson, J.P. and Ganiats, T.G. (1993) The Quality of Well Being scale: rationale for a single quality of life index. In: *Quality of Life Assessment: Key Issues in the 1990s*. (Walker, S.R. and Rosser, R.M. eds) Dordrecht/Boston/London: Academic Publishers.

Kasperson, R.E. (1992) The social amplification of risk: progress in developing an integrative framework. In: *Social Theories of Risk*. (Krimsky, S. and Golding, D. eds), pp.153-178. Westport, CT: Praeger.

Kendall, L. (1997) *Asthma and Air Quality. A Report to the Department of Health*. National Foundation for Educational Research.

Kind, P. (1989) *The Design and Construction of Quality of Life Measures: Discussion Paper No 43*. York: Centre for Health Economics.

Mullen, P. (1998) Priority Setting in Health Care: Techniques and Pitfalls. In: *Managing Health Care under Resource Constraints*. (Kastelein, A. *et al*, eds). pp.105-133. EURO, Eindhoven.

Okrent, D. and Pidgeon, N.F. (1998) Risk assessment versus risk perception. *Reliability Engineering and Systems Safety* **59**: 1-159.

Patrick, D.L., Bush, J.W. and Chen, M.M. (1973) Methods for measuring levels of well-being for a health status index. *Health Services Res.* **8**: 228-245.

Pidgeon, N.F. and Beattie, J. (1997). The psychology of risk and uncertainty. In: *Handbook of Environmental Risk Assessment and Management*. (Calow, P. ed), pp.289-318. Oxford: Blackwell Science.

Pidgeon, N.F., Hood, C., Jones, D. and Turner, B.A. (1992) Risk Perception. In: *Risk: Analysis, Perception and Management*. London: The Royal Society.

Renn, O. (1998) The role of risk perception for risk management. *Reliability Engineering and System Safety* **59**: 49-62.

Royal Commission on Environmental Pollution (1998). *Twenty-First Report. Setting Environmental Standards*. London: The Stationery Office.

Ryan, M. (1996a) *Using consumer preferences in health care decision-making: the application of conjoint analysis*. London: Office of Health Economics.

Ryan, M. (1996b) *Establishing the convergent validity of willingness to pay and conjoint analysis for eliciting preferences*. Aberdeen University: Health Economics Research Unit.

Slovic, P., Fischhoff, B. and Lichtenstein, S. (1980) Facts and fears: understanding perceived risk. In: *Societal Risk Assessment*. (Schwing, R. and Albers, W.A. eds). pp.181-124. New York: Plenum.

Stern, P.C. and Fineberg, H.V. (1996) *Understanding Risk: Informing Decisions in a Democratic Society*. Washington, DC: National Academy Press.

Tengs, T.O., Adams, M.E., Pliskin, J.S., Safran, D.G., Siegel, J.E., Weinstein, M.C. and Graham, J.D. (1995) Five-hundred life-saving interventions and their cost-effectiveness. *Risk Anal.* **15**: 369-390.

United States Environmental Protection Agency (1996). *Safe Drinking Water Act*. Washington, DC: US Environmental Protection Agency.

Viscusi, W.K., Magat, W.A. and Huber, J. (1991) Pricing environmental health risks: survey assessments of risk-risk and risk-dollar trade-offs for chronic bronchitis. *J. Environ. Econom. Manage.* **21**: 32-51.

Watson, S.R. and Buede, D.M. (1987) *Decision Analysis: The Principles and Practice of Decision Analysis*. Cambridge: Cambridge University Press.

Williams, A. (1997a) The Measurement and Valuation of Health: a Chronicle. In: *Being Reasonable About the Economics of Health*. (Culyer, A.J. and Maynard, A eds). Cheltenham, UK; Lyme, NH: Edward Elgar. pp. 136-175.

Williams, A. (1997b) Intergenerational equity: an exploration of the 'fair innings' argument. *Health Economics* **6**: 117-132.

ANNEX 3A

Quality of Life Measures

There are a variety of measures of health-related quality of life or health status available. This Annex covers two measures which are referred to in more detail in this report. These are the EQ5D scale, a utility index developed in recent years in the UK by the EuroQol group and the Quality of Well Being scale, a utility index with additional symptom scores widely used in the US. The EQ5D and the QWB both use scales from 1 for full health to 0 for a state equivalent to death. The questionnaires for each of these measures are given below. The EQ5D and the QWB questionnaires include the weightings for the various quality of life states which will be referred to in Chapter 6. The EQ5D weightings refer to 10 years in the relevant quality of life state and are derived from a survey of a representative sample of the general population (Williams, 1997). Weightings for 1 month and 1 year are also available (Measurement and Valuation of Health Group, 1995).

References

Measurement and Valuation of Health Group (1995). *The Measurement and Valuation of Health*, York: University of York, Centre for Health Economics.

Williams, A. (1997) The Measurement and Valuation of Health: a Chronicle. In: *Being Reasonable About the Economics of Health*. (Culyer, A.J. and Maynard, A. eds). Cheltenham, UK; Lyme, NH: Edward Elgar. pp. 136-175.

The EQ5D Scale

Your own health state today

By placing a tick in one box in each group below, please indicate which statement best describes your own health state today.

Do not tick more than one box in each group.

Mobility

I have no problems in walking about ☐

I have some problems in walking about ☐

I am confined to bed ☐

Self-Care

I have no problems with self-care ☐

I have some problems washing and dressing myself ☐

I am unable to wash or dress myself ☐

Usual Activities (eg. work, study, housework, family or leisure activities)

I have no problems with performing my usual activities ☐

I have some problems with performing my usual activities ☐

I am unable to perform my usual activities ☐

Pain/Discomfort

I have no pain or discomfort ☐

I have moderate pain or discomfort ☐

I have extreme pain or discomfort ☐

Anxiety/Depression

I am not anxious or depressed ☐

I am moderately anxious or depressed ☐

I am extremely anxious or depressed ☐

Your own health state today

To help people say how good or bad a health state is, we have drawn a scale (rather like a thermometer) on which the best state you can imagine is marked 100 and the worst state you can imagine is marked 0.

We would like you to indicate on this scale how good or bad your own health is today, in your opinion. Please do this by drawing a line from the box below to whichever point on the scale indicates how good or bad your health state is.

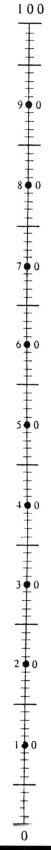

Scoring EQ-5D Health States

EQ-5D health states, defined by the health state classification, may be converted to a score using the tables of values shown on these two pages. The values for the 243 possible EuroQol health states plus unconsciousness, are based on data collected from a representative survey of the UK general public.

Health state	Score	Health state	Score	Health state	Score	Health state	Score
1 1 1 1 1	1.00	1 2 3 2 1	0.33	2 1 2 3 1	0.16	2 3 2 1 1	0.33
1 1 1 1 2	0.85	1 2 3 2 2	0.26	2 1 2 3 2	0.09	2 3 2 1 2	0.26
1 1 1 1 3	0.41	1 2 3 2 3	0.09	2 1 2 3 3	-0.08	2 3 2 1 3	0.10
1 1 1 2 1	0.80	1 2 3 3 1	0.07	2 1 3 1 1	0.49	2 3 2 2 1	0.21
1 1 1 2 2	0.73	1 2 3 3 2	0.00	2 1 3 1 2	0.42	2 3 2 2 2	0.14
1 1 1 2 3	0.29	1 2 3 3 3	-0.17	2 1 3 1 3	0.25	2 3 2 2 3	-0.03
1 1 1 3 1	0.26	1 3 1 1 1	0.44	2 1 3 2 1	0.36	2 3 2 3 1	-0.05
1 1 1 3 2	0.19	1 3 1 1 2	0.37	2 1 3 2 2	0.29	2 3 2 3 2	-0.13
1 1 1 3 3	0.03	1 3 1 1 3	0.20	2 1 3 2 3	0.13	2 3 2 3 3	-0.29
1 1 2 1 1	0.88	1 3 1 2 1	0.31	2 1 3 3 1	0.10	2 3 3 1 1	0.27
1 1 2 1 2	0.81	1 3 1 2 2	0.24	2 1 3 3 2	0.03	2 3 3 1 2	0.20
1 1 2 1 3	0.38	1 3 1 2 3	0.08	2 1 3 3 3	-0.13	2 3 3 1 3	0.04
1 1 2 2 1	0.76	1 3 1 3 1	0.05	2 2 1 1 1	0.75	2 3 3 2 1	0.15
1 1 2 2 2	0.69	1 3 1 3 2	-0.02	2 2 1 1 2	0.68	2 3 3 2 2	0.08
1 1 2 2 3	0.25	1 3 1 3 3	-0.19	2 2 1 1 3	0.24	2 3 3 2 3	-0.09
1 1 2 3 1	0.23	1 3 2 1 1	0.40	2 2 1 2 1	0.62	2 3 3 3 1	-0.11
1 1 2 3 2	0.16	1 3 2 1 2	0.33	2 2 1 2 2	0.55	2 3 3 3 2	-0.18
1 1 2 3 3	-0.01	1 3 2 1 3	0.16	2 2 1 2 3	0.12	2 3 3 3 3	-0.35
1 1 3 1 1	0.56	1 3 2 2 1	0.28	2 2 1 3 1	0.09	3 1 1 1 1	0.34
1 1 3 1 2	0.49	1 3 2 2 2	0.21	2 2 1 3 2	0.02	3 1 1 1 2	0.27
1 1 3 1 3	0.32	1 3 2 2 3	0.04	2 2 1 3 3	-0.14	3 1 1 1 3	0.10
1 1 3 2 1	0.43	1 3 2 3 1	0.01	2 2 2 1 1	0.71	3 1 1 2 1	0.21
1 1 3 2 2	0.36	1 3 2 3 2	-0.06	2 2 2 1 2	0.64	3 1 1 2 2	0.14
1 1 3 2 3	0.20	1 3 2 3 3	-0.22	2 2 2 1 3	0.21	3 1 1 2 3	-0.02
1 1 3 3 1	0.17	1 3 3 1 1	0.34	2 2 2 2 1	0.59	3 1 1 3 1	-0.05
1 1 3 3 2	0.10	1 3 3 1 2	0.27	2 2 2 2 2	0.52	3 1 1 3 2	-0.12
1 1 3 3 3	-0.07	1 3 3 1 3	0.11	2 2 2 2 3	0.08	3 1 1 3 3	-0.29
1 2 1 1 1	0.82	1 3 3 2 1	0.22	2 2 2 3 1	0.06	3 1 2 1 1	0.30
1 2 1 1 2	0.74	1 3 3 2 2	0.15	2 2 2 3 2	-0.02	3 1 2 1 2	0.23
1 2 1 1 3	0.31	1 3 3 2 3	-0.02	2 2 2 3 3	-0.18	3 1 2 1 3	0.06
1 2 1 2 1	0.69	1 3 3 3 1	-0.04	2 2 3 1 1	0.38	3 1 2 2 1	0.18
1 2 1 2 2	0.62	1 3 3 3 2	-0.11	2 2 3 1 2	0.31	3 1 2 2 2	0.11
1 2 1 2 3	0.19	1 3 3 3 3	-0.28	2 2 3 1 3	0.15	3 1 2 2 3	-0.06
1 2 1 3 1	0.16	2 1 1 1 1	0.85	2 2 3 2 1	0.26	3 1 2 3 1	-0.09
1 2 1 3 2	0.09	2 1 1 1 2	0.78	2 2 3 2 2	0.19	3 1 2 3 2	-0.16
1 2 1 3 3	-0.08	2 1 1 1 3	0.35	2 2 3 2 3	0.02	3 1 2 3 3	-0.32
1 2 2 1 1	0.78	2 1 1 2 1	0.73	2 2 3 3 1	0.00	3 1 3 1 1	0.24
1 2 2 1 2	0.71	2 1 1 2 2	0.66	2 2 3 3 2	-0.07	3 1 3 1 2	0.17
1 2 2 1 3	0.27	2 1 1 2 3	0.22	2 2 3 3 3	-0.24	3 1 3 1 3	0.01
1 2 2 2 1	0.66	2 1 1 3 1	0.20	2 3 1 1 1	0.37	3 1 3 2 1	0.12
1 2 2 2 2	0.59	2 1 1 3 2	0.12	2 3 1 1 2	0.30	3 1 3 2 2	0.05
1 2 2 2 3	0.15	2 1 1 3 3	-0.04	2 3 1 1 3	0.13	3 1 3 2 3	-0.12
1 2 2 3 1	0.12	2 1 2 1 1	0.81	2 3 1 2 1	0.24	3 1 3 3 1	-0.14
1 2 2 3 2	0.05	2 1 2 1 2	0.74	2 3 1 2 2	0.17	3 1 3 3 2	-0.21
1 2 2 3 3	-0.11	2 1 2 1 3	0.31	2 3 1 2 3	0.01	3 1 3 3 3	-0.38
1 2 3 1 1	0.45	2 1 2 2 1	0.69	2 3 1 3 1	-0.02	3 2 1 1 1	0.23
1 2 3 1 2	0.38	2 1 2 2 2	0.62	2 3 1 3 2	-0.09	3 2 1 1 2	0.16
1 2 3 1 3	0.22	2 1 2 2 3	0.19	2 3 1 3 3	-0.25	3 2 1 1 3	0.00

Health state	Score	Health state	Score	Health state	Score	Health state	Score
3 2 1 2 1	0.11	3 2 2 3 2	-0.26	3 3 1 1 3	-0.11	3 3 2 3 1	-0.30
3 2 1 2 2	0.04	3 2 2 3 3	-0.43	3 3 1 2 1	0.00	3 3 2 3 2	-0.37
3 2 1 2 3	-0.13	3 2 3 1 1	0.14	3 3 1 2 2	-0.07	3 3 2 3 3	-0.54
3 2 1 3 1	-0.15	3 2 3 1 2	0.07	3 3 1 2 3	-0.24	3 3 3 1 1	0.03
3 2 1 3 2	-0.22	3 2 3 1 3	-0.10	3 3 1 3 1	-0.26	3 3 3 1 2	-0.04
3 2 1 3 3	-0.39	3 2 3 2 1	0.02	3 3 1 3 2	-0.33	3 3 3 1 3	-0.21
3 2 2 1 1	0.20	3 2 3 2 2	-0.06	3 3 1 3 3	-0.50	3 3 3 2 1	-0.09
3 2 2 1 2	0.13	3 2 3 2 3	-0.22	3 3 2 1 1	0.09	3 3 3 2 2	-0.17
3 2 2 1 3	-0.04	3 2 3 3 1	-0.25	3 3 2 1 2	0.02	3 3 3 2 3	-0.33
3 2 2 2 1	0.07	3 2 3 3 2	-0.32	3 3 2 1 3	-0.15	3 3 3 3 1	-0.36
3 2 2 2 2	0.00	3 2 3 3 3	-0.48	3 3 2 2 1	-0.04	3 3 3 3 2	-0.43
3 2 2 2 3	-0.16	3 3 1 1 1	0.12	3 3 2 2 2	-0.11	3 3 3 3 3	-0.59
3 2 2 3 1	-1.19	3 3 1 1 2	0.05	3 3 2 2 3	-0.27	Unconscious (-0.40)	

UK A1 Version

To score a health state, simply read off the corresponding value and record it in the box on the front of the questionnaire.

Table 3A.1 Quality of Well-being/General Health Policy Model: elements and calculating formulas (function scales, with step definitions and calculating weights)

Step No.	Step Definition	Weight
	Mobility Scale (MOB)	
5	No limitations for health reasons	-.000
4	Did not drive a car, health related; did not ride in a car as usual for age (younger than 15 yr), health related, and/or did not use public transportation, health related; or had or would have used more help than usual for age to use public transportation, health related	-.062
2	In hospital, health related	-.090
	Phsyical Activity Scale (PAC)	
4	No limitations for health reasons	-.000
3	In wheelchair, moved or controlled movement of wheelchair without help from someone else; or had trouble or did not try to lift, stoop, bend over, or use stairs or inclines, health related; and/or limped, used a cane, crutches, or walker, health related; and/or had any other physical limitation in walking, or did not try to walk as far as or as fast as other the same age are able, health related	-.060
1	In wheelchair, did not move or control the movement of wheelchair without help from someone else, or in bed, chair, or couch for most or all of the day, health related	-.077
	Social Activity Scale (SAC)	
5	No limitations for health reasons	-.000
4	Limited in other (e.g. recreational) role activity, health related	-.061
3	Limited in major (primary) role activity, health related	-.061
2	Performed no major role activity, health related, but did perform self-care activities	-.061
1	Performed no major role activity, health related, and did not perform or had more help than usual in performance of one or more self-care activities, health related	-.106

Calculating Formulas

Formula 1. Point-in-time well-being score for an individual (W):

$$W = 1 + (CPXwt) + (MOBwt) + (PACwt) + (SACwt)$$

where "wt" is the preference-weighted measure for each factor and CPX is Symptom/Problem complex. For example, the W score for a person with the following description profile may be calculated for one day as:

CPX-11	Cough, wheezing or shortness of breath, with or without fever, chills, or aching all over	-.257
MOB-5	No limitations	-.000
PAC-1	In bed, chair, or couch for most or all of the day, health related	-.077
SAC-2	Performed no major role activity, health related, but did perform self care	-0.61

$$W = 1 + (-.257) + (-.000) + (-.077) + (-.061) = 0.605$$

Formula 2. Well-years (WY) as an output measure:

$$WY = [\text{No. of persons} \times (1.0 + CPXwt + MOBwt + PACwt + SACwt) \times \text{Time}]$$

CPX No.	CPX Description	Weights
1	Death (not on respondent's card)	-.727
2	Loss of consciousness such as seizure (fits), fainting, or coma (out cold or knocked out)	-.407
3	Burn over large areas of face, body, arms or legs	-.387
4	Pain, bleeding, itching or discharge (drainage) from sexual organs – does not include normal menstrual (monthly) bleeding	-.349
5	Trouble learning, remembering, or thinking clearly	-.340
6	Any combination of one or more hands, feet, arms, or legs either missing, deformed (crooked), paralysed (unable to move), or broken – includes wearing artifical limbs or braces	-.333
7	Pain, stiffness, weakness, numbness, or other discomfort in chest, stomach (including hernia or rupture), side neck, back, hips or any joints or hands, feet, arms or legs	-.299
8	Pain, burning, bleeding, itching, or other difficulty with rectum, bowel movements, or urination (passing water)	-.292
9	Sick or upset stomach, vomiting or loose bowel movement, with or without chills, or aching all over	-.290
10	General tiredness, weakness, or weight loss	-.259
11	Cough, wheezing, or shortness of breath, with or without fever, chills, or aching all over	-.257
12	Spells of feeling, upset, being depressed, or of crying	-.257
13	Headache, or dizziness, or ringing in ears, or spells of feeling hot, nervous or shaky	-.244
14	Burning or itching rash on large areas of face, body, arms or legs	-.240
15	Trouble talking, such as lisp, stuttering, hoarseness, or being unable to speak	-.237
16	Pain or discomfort in one or both eyes (such as burning or itching) or any trouble seeing after correction	-.230
17	Overweight for age and height or skin defect of face, body, arms, or legs, such as scars, pimples, warts, bruises or changes in colour	-.188
18	Pain in ear, tooth, jaw, throat, lips, tongue; several missing or crooked permanent teeth – includes wearing bridges or false teeth; stuffy, runny nose; or any trouble hearing – includes wearing a hearing aid	-.170
19	Taking medication or staying on a prescribed diet for health reasons	-.144
20	Wore eyeglasses or contact lenses	-.101
21	Breathing smog or unpleasant air	-.101
22	No symptoms or problem (not on respondent's card)	-.000
23	Standard symptom/problem	-.257
X24	Trouble sleeping	-.257
X25	Intoxication	-.257
X26	Problems with sexual interest or performance	-.257
X27	Excessive worry or anxiety	-.257

Note: X (i.e. X24) indicates a standardized weight and is used because preference studies have not been completed

ANNEX 3B
Multiple Criteria Decision Analysis

Introduction

1 A brief summary of the principles of this method was given in Chapter 3. This annex considers these in more detail and discusses how multiple criteria decision analysis (MCDA) might be used in the air pollution context.

2 The essence of multiple criteria (or "multiple attribute") models is the recognition that alternative policies are often better or worse on many dimensions. This is so even for a simple choice such as the purchase of a bicycle, where relevant criteria may include initial price, likely depreciation, weight, comfort, robustness and so on. In the present context, measures to improve air quality will generally have many other effects, direct and indirect. Multiple criteria methods provide a framework for trading-off benefits on different dimensions. The key is to establish a set of common criteria against which all the alternative choices are to be evaluated. Rather than using a single scale throughout, as with monetary valuation, the idea is to measure how well each alternative performs against each separate criterion, then prioritise or "weight" the criteria to arrive at an overall judgement. Multiple criteria methods form a wide "family" of approaches.

3 In principle, multiple criteria models have some appeal in analysing issues involving health and safety, particularly since many types of cost and benefit are often involved. Consider, for example, a proposal to lower the national speed limit on roads. This will have an impact on accidental deaths and injuries, while lengthening travel times (in itself likely to have many indirect effects, perhaps including a switch by some from road to rail, but also greater costs to industry). As for less tangible effects, many people value the enjoyment of driving fast. Others would put higher value on the feeling of safety produced by everyone driving slower. The proposal should also reduce levels of at least some pollutants and this should further reduce deaths and ill health, effects which (as has been argued at length) are not valued in the same way as accidental death and injury. Meanwhile a decrease in pollution would have other environmental benefits.

4 Costing risks to health is, thus, intertangled with many other problems of measurement. Rather than trying to relate everything to a single scale such as monetary value, it might seem more realistic to measure these disparate costs and benefits on different scales. Having done this, it will still be necessary to make trade-offs. (By contrast, measurement on a single scale means that costs and benefits are directly comparable, and can simply be added together.)

MCDA in practice

5 While methods such as WTP and conjoint analysis surveys are principally ways to investigate how people in general value different "goods", MCDA is primarily a means of assisting decisions. Decision analysis itself deals with decisions that involve risks - in the sense that the results of each choice are governed by probabilities rather than being certain. *Multiple criteria decision analysis* allows for the extra point that there may be several, at least partly competing objectives. Medical decisions, for example, may involve balancing expected survival against likely quality of life (itself, a multi-dimensional matter as recognised in the construction of measures such as quality-adjusted life years).

6   The basic steps involved in setting up an MCDA model are outlined in the main text. In practice, the process will usually be far from linear, with earlier steps revisited as the model is built up. The use of flexible, interactive software allows models to be displayed, analysed and changed rapidly. In particular, it is easy to carry out wide-ranging *sensitivity analysis*, in which judgements about scores or criteria weights can be revised to see when and how the "best" course of action would change. This helps show which are the most critical areas of uncertainty which may require further research.

7   MCDA has a growing track record in helping decisions in both public and private sectors (Watson and Buede, 1987; Belton, 1990), some of which involve balancing economic criteria with various health risks. Of particular interest is the "risk ranking" methodology (Chicken and Hayns, 1989) developed for the evaluation of large projects and currently under consideration by HSE (Beattie *et al*, 1998). MCDA has also been developed as a way of addressing the management of individual health and safety decisions (Keeney and von Winterfeldt, 1991). An example of its use in the assessment of options for the management of solid radioactive waste is given in "Policy Appraisal and the Environment" (Department of the Environment, 1991).

8   In the air pollution context, MCDA, could be used to compare a variety of pollution control options (traffic restrictions, fuel composition, vehicle standards, industrial controls, no further action etc.) according to a series of criteria with different relative weights (e.g., costs, health impacts, inconvenience to motorists, effect on rural isolation, effect on the environment etc.). This would assist in identifying the best option or combination of options. Alternatively, it could be used to express the health benefits of each option in terms of an overall score. In this case the criteria would be a variety of different health outcomes (e.g., deaths brought forward in the elderly, deaths brought forward in adults, respiratory hospital admissions, cardiovascular hospital admissions, increased asthma symptoms) which could be weighted by relative importance. The "scores" would represent the extent to which an option for pollution reduction would impact on each of these. These would then be multiplied by the weight assigned to each health outcome and the weighted scores added up to an overall score. The weights could be derived from a panel of experts or a sample of the public. This would require a specific study and it is difficult, in the light of current knowledge, to infer what sort of weights might be given.

9   In considering the possible use of MCDA, it is essential to bear in mind the distinction made in the main text, concerning the question of *whose values are being sought*.

One ideal is to seek the values of a wide population - those who will be affected by a risk - and then use these values as an input to decisions. The aim is to inform policy-makers about the values they *should* place on health risks if they are to reflect the wishes of the public.

The opposite ideal is to start with the values of those making the decisions. The aim is to structure and clarify these, so helping decision-makers choose policies that best satisfy them.

Willingness to pay surveys, for example, start from the first ideal, while as a decision-aiding method MCDA quite properly starts from the second. However, the scope of MCDA can be expanded by widening participation in the modelling exercise to include not only the responsible policy-makers and their advisors, but independent "experts" on particular topics, representatives of stakeholder groups, and so on (Ives and Footitt, 1996).

10 If MCDA is used to capture the views of a wider public by structuring the proceedings of "focus groups", "citizen juries" and the like, the gap between its use and that of (say) Conjoint Analysis narrows. The choice between single and multiple criteria models may also be seen as a pragmatic one. Decisions about risks to public health necessarily involve eventual trade-offs with different types of benefits. The question is whether to make generic trade-offs (e.g., through the "value of prevention of a statistical fatality") so as to translate all costs and benefits into monetary terms *throughout* the analysis, or to evaluate choices on many criteria before making a final, and explicit trade-off. Given the complexity of the issues involved, the most fruitful approach in the longer term may well be to use different approaches alongside each other, and to cross-check the results obtained.

References

Belton, V. (1990) Multiple criteria decision analysis - practically the only way to choose. In: *Operational Research Society Conference Tutorial Papers.* (Hendy LC, Eglese, R. eds). Birmingham: Operational Research Society.

Beattie, J., Chilton, S., Cookson, R., Covey, J., Hopkins, L., Jones-Lee, M., Loomes, G., Pidgeon, N.F., Robinson, A. and Spencer, A. (1998) *Valuing Health and Safety Controls: A Literature Review.* London: HSE Books.

Chicken, J.C. and Hayns, M.R. (1989) *The Risk Ranking Technique in Decision-Making.* Oxford: Pergamon Press.

Department of the Environment (1991). *Policy Appraisal and the Environment: A Guide for Government Departments.* London: HMSO.

Ives, D.P. and Footitt, A.J. (1996) *Risk Ranking. Final report submitted to HSE Research Strategy Unit RSU 3444/R71.011.* University of East Anglia: Centre for Environmental and Risk Management.

Keeney, R.L. and von Winterfeldt, D. (1991) A Prescriptive risk framework for individual health and safety decisions. *Risk Anal.* **11**: 523-533.

Watson, S.R. and Buede, D.M. (1987) *Decision Analysis: The Principles and Practice of Decision Analysis.* Cambridge: Cambridge University Press.

Chapter 4

Benefits of Lower Mortality Risks

Introduction

4.1 This chapter assesses the ways in which the benefits of lower mortality risks can be measured and valued. It also compares ways in which these benefits could be incorporated into economic approaches to decision-making for use in assessing potential health care and environmental policies in the UK.

4.2 We are very aware that when the UK's resources are devoted to measures to reduce pollution they are not then available for other uses - e.g., for personal consumption, other public sector programmes, or health care. As previously noted, we therefore feel that it is highly relevant, wherever practicable, to try to value the health benefits brought about by lower levels of air pollution, so that full appraisal can be carried out to show that resources under the government's control are devoted to the uses of greatest value to UK citizens. The benefits of reduced mortality risks are likely to be the most important benefits, and so we believe it is worth dedicating significant effort to estimating their value.

4.3 Chapter 2 explains the effects of air pollution on mortality risks. In essence the epidemiological research has characterised three mortality patterns attributable to air pollution.

- *Acute risks*: deaths in susceptible individuals may occur following short-term (e.g., day-to-day) fluctuations in pollution levels.

- *Chronic risks*: deaths may occur following the development of chronic disease and progressive deterioration of health, to which long-term exposure to pollution may contribute.

- *Latent risks*: a form of chronic risk in which deaths may occur many years after initial exposure to pollution, but without many years of progressive deterioration of health.

The evidence is strongest for effects on acute mortality.

Valuing benefits: whose values?

4.4 The benefits that need to be appraised are those accruing to all potential beneficiaries. As we explain below, large parts of the population may attach value to reducing pollution.

4.5 In the case of the acute risks, those who die are likely to be, by the time of their death, already in very poor health. This means that they will have a short life expectancy, whatever the level of air pollution. However, in assessing the population which is interested in reducing these risks it is relevant to consider the position not just of those who are already in very poor health, but also of anyone who might become so, and thus be vulnerable to air pollution, in the period before control measures take effect. There is likely to be much uncertainty as to which current members of the UK population may eventually fall into the high risk groups who are most vulnerable to the effects of air pollution and would benefit most from its control. This uncertainty is likely to increase as lead-in times increase. Moreover, there are no *a priori* reasons to believe that there will be significant differences in preferences for risk reduction between those who will become vulnerable and those who will not. Thus, the preferences of the general population may be the most representative of those who would eventually benefit.

4.6 Those potentially at chronic risk will gain direct benefits through reduction of cumulative exposure. The evidence for effects on chronic risks is less certain than for acute risks, and there is less understanding of whether some groups are more at risk than others.

4.7 Those potentially at latent risk will gain direct benefits through reduction of carcinogenic air pollutants. Although there probably is some variation in individual susceptibility to carcinogens, there are currently no clearly identifiable sub-groups at increased or decreased risk.

4.8 Paragraphs 4.5 to 4.7 above identify the interests of the direct beneficiaries[1]. Unfortunately, there are major gaps in our knowledge about the quantitative health impact of the chronic and latent risks and who might benefit most from their reduction.

Mortality effects

4.9 Chapter 2 reports what it is currently possible to say about the numbers of deaths, based on the COMEAP report "*Quantification of the Effects of Air Pollution on Health in the UK*" (Department of Health, 1998). For example, it is estimated that each year in GB as many as 12,500 deaths from respiratory disease are brought forward by acute exposure to ozone air pollution.

4.10 Although such deaths are premature, the loss of life expectancy involved may not be large. It might range from one month to one year. With a loss of one year of life expectancy per death, for example, this would imply that ozone air pollution was responsible for an aggregate annual loss of 12,500 life-years. On this basis, the annual mortality effects of ozone could be described in terms of either 12,500 premature deaths, or 12,500 life-years lost. If the loss of life expectancy per death was only one month, however, the sum would be one twelfth of this last figure - about 1,040 life-years lost.

Economic approaches to health care policy-making

4.11 Chapter 3 has discussed different approaches to decision-making in the context of pollution reduction. Cost-benefit analysis (CBA) seeks to maximise overall net benefits through balancing the benefits and the costs of policies, while cost-effectiveness analysis (CEA) seeks to maximise effectiveness within the constraint of a pre-determined budget. CEA is now used extensively in health care, where it typically employs standardised measures of mortality and/or morbidity improvement. For example, given the budget, the objective might be to minimise the number of life-years lost.

4.12 As described in Chapter 3, a particular variant being widely adopted in health care policy-making uses a measure of effectiveness that joins mortality and morbidity outcomes together. In this type of CEA the aim is to maximise life expectancy weighted by changes in health-related quality of life - i.e., the number of "quality-adjusted life years" that can be added to people's lives - with available resources.

4.13 Health care policy-making now aims to consider systematically the extent to which different policies affect health - for example whether the health gains from using resources in one way would be greater than the health gains from using them in some other way. Assessing quality-adjusted life years gained (and costs) can greatly help with such systematic appraisal.

---

[1] Those who are not potentially direct beneficiaries may still feel some benefit from reduced air pollution through their feelings about the effects on their relatives, or through altruism towards their fellow citizens (see also paragraph 3.26). Whether such altruism should be seen as a reason for increasing the estimated aggregate value of pollution reduction will depend on whether it is pure altruism or paternalistic altruism. Pure altruists will attach value to whatever their fellow citizens value, to the extent that they value it - but will attach neither more nor less value than that, because to do so would distort or double-count the values of their fellow citizens. Paternalistic altruists will attach greater value to certain "good things" which they feel would be of benefit to their fellow citizens and which they feel their fellow citizens under-value. Paternalistic altruists are willing to pay more for these good things for others, so as to "correct" the "under-valuation". For a more rigorous treatment of these issues see Jones-Lee (1992).

4.14 It is, therefore, of interest to know how many quality-adjusted life years would be added to people's lives through reducing air pollution, and how this would compare with the quality-adjusted life years to be obtained through other health policies. Given available resources, policies can then be ranked and selected so as to maximise gains in life expectancy weighted by changes in quality of life.

Quantity and quality of life lost due to premature deaths

4.15 Unfortunately, there is a lack of evidence about the health-related quality of life of people who are at significant risk of premature death from air pollution. Inferences about those who die from a worsening of respiratory disease triggered by air pollution might be drawn from studies of patients who are seriously ill with respiratory disease in general. While it is not known whether such patients are representative of those whose deaths are brought forward by air pollution, it seems reasonable to suspect that there will be many similarities. An example is the report by Osman *et al* (1997) who studied quality of life in patients admitted to hospital in Aberdeen with chronic obstructive pulmonary disease (COPD). The fate of the patients (readmissions, deaths, etc.) was then followed over the next 12 months. The patients who died within this 12 month period had had substantially impaired quality of life at the beginning of the period - typically falling into the worst half of the particular scale used to measure quality of life in this study (the St George's Respiratory Questionnaire).

4.16 A more recent study (Ayres *et al*, personal communication) has measured the health of COPD patients in Birmingham. In this study, health-related quality of life was measured using the EQ5D instrument. The patients themselves rated their own health on the five dimensions of EQ5D: mobility, ability to perform self-care, ability to perform usual activities, pain/discomfort, and anxiety/depression. (On this scale 1 represents the best of health and 0 is as bad as being dead - see Chapter 3.) Again it was found that the patients who died during the index admission had had substantially impaired quality of life before being hospitalised: on this scale, these COPD patients rated their health-related quality of life at about 0.4 on average ($\pm$ 1 Standard Deviation (SD) from 0.2 to 0.7)[2]. Our view is that this represents the best estimate that can currently be made of the quality of life of people with COPD whose deaths are brought forward by air pollution.

4.17 A key issue is the effect that reducing air pollution would have. While it is predicted that improvements in air quality would reduce the number of premature deaths we really need to know too whether these people's quality of life would be improved at the same time as their lives would be extended. Informed clinical judgement suggests, however, that even if reductions in risks were to postpone their deaths, their underlying chronic ill-health would not be improved.

4.18 Given the limited evidence, it is not yet possible to give good estimates of the number of quality-adjusted life years lost owing to deaths brought forward by air pollution. While around 12,500 premature deaths from respiratory disease may be associated with ozone pollution per year, it is not possible accurately to say how this would translate into the number of quality-adjusted life years lost. It seems reasonable to suppose that it would be less than this. The best estimate that we can make is that the loss of life expectancy would range from one month to one year, and the factor that should be applied for health-related quality of life is 0.4 ($\pm$ 1 SD 0.2 to 0.7). The quality-adjusted life years that would be lost per death brought forward would, therefore, be between 0.03 (0.015 - 0.06) and 0.4 (0.2 - 0.7) for a loss of one month and one year, respectively. Thus, with a year of lost life expectancy per death, we would estimate that about 5,000 (2,500 - 8,750) quality-adjusted life years would be lost due to ozone pollution mortality. Similarly, the quality-adjusted life years lost due to PM_{10}-induced premature mortality may be around 3,200 (1,600 - 5,700). The quality-adjusted life years lost due to sulphur dioxide-induced premature mortality may be around 1,400 (700 - 2,500). If the loss of life expectancy per death was only one month, however, the number of quality-adjusted life years lost would be one twelfth of these figures: about 420 (210 - 730) for ozone, 270 (135 - 475) for PM_{10}, and 120 (60 - 205) for sulphur dioxide.

---

[2] While the observed standard deviation is not the measure usually used to indicate the range of values within which we are confident the true value should lie, in this case it is, unfortunately, the only measure available to us (see paragraph 7.54).

4.19 The above estimates of quality-adjusted life years lost relate to deaths from respiratory disease. For air pollution-induced premature deaths from cardiovascular disease there is even less evidence. It is thought likely that there will be greater variation in the state of health of such people before their death - some of them may have chronic heart disease that is painful and/or limits what they can do, while others may notice no symptoms and may lead a normal life. Further research is needed to assess how air pollution affects cardiovascular health.

4.20 Deaths from cardiovascular disease occur mainly in those aged 65 and over. We do not actually know whether the health of people who suffer premature death from air pollution-induced cardiovascular disease is better or worse than that of the general population of the same age. Instead, for illustration, we show the implications of a quality of life rating equivalent to that of the general elderly population. We note that the *1996 Health Survey for England* (Department of Health, 1998) assessed health-related quality of life in the general population using, *inter alia*, the EQ5D measure. It found the average rating, on the 0-1 scale, to be 0.78 for people aged 65-74 and 0.73 for those aged 75 or older - and a population-weighted average figure for the whole 65+ age group would be 0.76.

4.21 For now we leave the issue of quality-adjusted life years. We do, though, return to this later in this Chapter and in Chapter 6. We turn next to the issue of valuing reductions in risk of premature mortality. In this we try to explain how we think the available valuation evidence should be interpreted for use in cost-benefit analysis.

Valuation of mortality risks

4.22 Accepted valuation principles ascribe value to anything for which people are willing to pay some amount[3]. It is not necessary for "payment" to be monetary - it simply refers to the idea of "giving something up", which would include time or effort just as much as cash payment. Conventionally this is known as the *willingness to pay* (WTP) approach. This approach is grounded in people's preferences and reflects the strengths of those preferences. It is, therefore, entirely appropriate for the purposes of endeavouring to put resources to their most highly valued uses.

4.23 The WTP approach can be used to assess the value which people put on reductions in risks. This is dealt with more fully in Chapter 3 (see in particular paragraph 3.30). Again, we stress that the issue is not about valuing an individual, known, life, but about the value that people themselves place on typically very small reductions in the risk they face. The collective values of risk reduction of a large group of people are typically aggregated up to the level where, statistically, one fewer premature fatality would be expected - and we term this the "VPF" for short[4].

Evidence

4.24 For our task it would have been helpful if empirical studies had already been carried out to try to assess the value that people put on reductions in risks from air pollution. Unfortunately, no work that directly addresses the questions that we want to ask has yet been undertaken.

4.25 There is, however, substantial evidence concerning the value that people attach to changes in mortality risks in other contexts. A considerable amount of good research, in many countries, has now been carried out to estimate the value people put on reducing risks of death in car crashes. Much research has also assessed how wages reflect risks in certain occupations. And there is, too, some evidence from people's willingness to spend money on safety, such as purchases of smoke alarms.

---

[3] One approach that was used a few decades ago was merely to calculate how much more would be produced or consumed by those at risk if they lived longer as a result of risks being reduced - the "Gross Output" approach. There are many faults with such a simplistic approach. A common method was to base such calculations on earnings - so for those not in work the figure calculated was zero. The approach was criticised by many - including economists - for being inconsistent with accepted principles for valuing economic goods. This approach is, therefore, not considered further here.

[4] As noted in Chapter 3, this concept has often been referred to in the literature as the "value of a statistical life" (see especially paragraph 3.30).

4.26 By now there are numerous examples of all three types of evidence, and of the empirical estimation procedures used to derive them. To report details of all of them would be beyond the scope of a report such as this. Suffice it to say that in several cases the studies concerned have yielded a VPF well in excess of £1 million[5].

The DETR's values for the prevention of fatal and non-fatal road casualties

4.27 The first UK government department to adopt the WTP approach to the valuation of preventing a fatality was the (then) Department of Transport (DoT) - now the Department of the Environment, Transport and the Regions (DETR). In particular, in 1988, following a comprehensive review of the literature[6] and careful consideration of the various issues involved, the DoT decided to abandon its former "gross output"-based valuation procedure. Instead it adopted a WTP-based value for the prevention of a road fatality of £500,000 (in 1987 prices) thereby effectively doubling the figure concerned. It is worth noting that the figure of £500,000 lay at the bottom end of the range of empirical estimates available in 1988 and the decision to select such a conservative estimate reflected the DoT's desire to exercise an element of caution in the face of a radical change of methodology. Since 1988 the original figure of £500,000 has been updated annually in line with inflation and the rate of growth of real income per capita.

4.28 More recently, on the basis of the findings of research carried out for the Department by the Universities of Newcastle upon Tyne and York, the DoT effected similar revisions of its valuation methodology for the prevention of serious and slight *non-fatal* road injuries.

4.29 In summary, in 1996 prices, the DETR's WTP-based values for the prevention of various severities of road injury currently stand at:[7]

Fatality	£847,580
Serious Injury	£ 96,620
Slight Injury	£ 7,480

4.30 While values of risk reduction from road safety are now firmly established, it seems clear to us that careful adaptation is necessary before these surveys can be used as the basis for figures applicable to the circumstances of air pollution. Hereafter, this chapter considers whether appropriate adjustments can be made. Our conclusion is that values derived in other contexts, notably road safety, can be used in deriving rough estimates of values to be applied in the air pollution context. We explain our reasons for, and methods of, making adjustments in some detail so that others can both understand what we have attempted to do and can be clear about whether or not they agree.

Application to the air pollution context

4.31 We believe that further consideration is needed before simply using values that were derived in other circumstances, since it is highly likely that WTP-based valuations of safety measures will be affected by a range of factors relevant to the air pollution context. Potentially significant factors include those set out in Table 4.1 on the following page.

---

[5] For a more detailed account of empirical work in this area, see for example, Jones-Lee, (1989), Ch.2, Viscusi (1993) and Beattie et al (1998).

[6] See Dalvi (1988)

[7] See *Highways Economics Note No. 1*, September 1997. Of the figure for a fatality, a small element - around £60,000 - is for lost work output net of consumption, and a smaller element is for medical costs. The largest element represents WTP.

Table 4.1 **Factors which may influence people's willingness to pay for avoiding particular risks**

(i)	Type of health effect (acute; chronic; latent)	e.g., people may dread a lingering death more than a sudden death
(ii)	Factors related to risk context (such as voluntariness; control; responsibility; uncertainty etc.)	e.g., people seem to regard involuntary risks over which they have no control, risks which are someone else's responsibility, and vague risks, as worse than others
(iii)	Futurity of health effect and discounting	e.g., effects which happen sooner are expected to be regarded as worse than those which happen later
(iv)	Age	e.g., people may attach particular value to life and health at certain ages
(v)	Remaining life expectancy	e.g., WTP is expected to be positively related to the number of years of life expectancy at risk (although not necessarily in direct proportion)
(vi)	Attitudes to risk	e.g., risk aversion is expected to affect willingness to trade wealth for risk; younger people may be less averse to risk
(vii)	State of health-related quality of life	e.g., people are expected to be keener to extend life in good health than life in poor health
(viii)	Level of exposure to risk	e.g., people may be keener to reduce a high risk by a set amount than a low risk by the same amount
(ix)	Wealth/income/socio-economic status	e.g., people with more wealth are likely to have a higher WTP to reduce a given risk than those with less, and there may be other differences between social groups

4.32 Of the factors in Table 4.1, the first three relate to the nature of the risk itself, while the fourth to ninth factors relate to the people exposed to the risk. We discuss below in some detail how the characteristics of both the risk and the people exposed to it are expected to affect preferences for risk reduction, and why these need to be considered before transferring a VPF from another context.

4.33 Of the published studies of the value that people put on safety improvements that reduce mortality risks (e.g., see Beattie *et al*, 1998), many relate to road safety and, of the remainder, most relate to occupational safety. The typical victim who dies from such risks is likely to be middle-aged and in an average state of health (for their age). For example, the average age at death of men who die in motor vehicle traffic accidents is 39.6, and for women it is 49.1 (Office of Population, Censuses and Surveys, 1994) and, at such an age, men would normally have a further life expectancy of about 36 years and women 32 years.

4.34 In the air pollution context, by contrast, a high proportion of those most affected are over 65 (see paragraphs 2.11-2.12). As discussed in paragraph 2.10, the extent of life expectancy lost is not accurately known, but it might range from less than 1 month to 1 year. Concomitantly, with such a short life expectancy, health would probably be significantly impaired (see paragraph 4.16). We consider that these characteristics are too different from those in other contexts to be able to apply the relevant WTP figures without adjustment. We therefore consider below the influence these characteristics might be expected to have on WTP.

4.35 The way we have approached this has been to develop a series of adjustments that will allow a starting point figure, or baseline value, to be modified by "multipliers" to produce a value that is tailored to any given profile of the factors listed in Table 4.1. However, given the paucity of empirical evidence concerning WTP-based values of preventing premature fatalities in the air pollution context, we stress that any precise set of adjustments must largely be speculative - although there is probably not too much difficulty in agreeing upon the qualitative *direction* of the impact of variations in most of the above factors upon the VPF.

The air pollution baseline VPF

4.36 The first step is to derive a baseline VPF allowing for the particular context of air pollution risks and applying to an average person (average age, health, wealth, etc.). It would be possible to conduct new surveys which would provide the basis for this figure but, as explained, there is no existing direct evidence for a pollution-based VPF. Consequently, we have had to rely on indirect evidence and inference.

4.37 A number of different means of deriving a baseline figure applicable in the context of air pollution risks have been examined. One approach is to use the methodology advocated in the NERA/CASPAR report (1998). The basic method is to apply air pollution risk "context multipliers" to the DETR roads VPF (some £0.8m, at 1996 prices, for the WTP element) on the grounds that account should be taken of public attitudes to the types of health effects and types of risk involved (see (i) and (ii) in Table 4.1). While people generally perceive road accident risks as largely voluntary, well-understood and relatively easy to control, air pollution risks are perceived as involuntary, poorly understood and not under individuals' control. Empirical evidence suggests that WTP for reductions in such involuntary risks can exceed that of voluntary risks by a factor of up to 2 or 3.[8]. Consequently, by factoring up the DETR VPF, we produce a pollution baseline VPF of around £2 million.

4.38 We have also looked at systematic reviews of other estimates of WTP to reduce the risk of a fatality. There are now many such studies from different contexts and using different methodologies. When studies that use the best and most rigorous methodologies are selected and the values they have derived are averaged, the results tend to yield figures for a VPF in the range £2-2.5 million. It is tempting to take comfort in this rough convergence with our first approach of factoring up the DETR figure. In fact, given the different assumptions and methodologies behind the different estimates, the convergence may be less significant than it appears. That said, it remains our view that the case for using a higher baseline figure than that used for road-related fatalities is very strong. Thus, we believe that it is reasonable to proceed on the basis of an air pollution baseline VPF of £2million. This baseline figure represents an average for people of all ages. We have no doubt that this figure needs to be checked against the results of new empirical work specifically set in the context of air pollution.

4.39 We believe that a baseline figure of this magnitude would be consistent with the approach adopted in other policy areas - such as transport - where explicit values are already placed on preventing fatalities, most notably for policies which prevent road deaths. However, while cross-programme consistency seems generally to be a desirable aim, we have to accept that different policy criteria have often been developed in different policy areas, with health care a generally distinct area of policy-making (see quality-adjusted life year discussion at paragraphs 4.71 to 4.76) so the point is by no means decisive.

[8.] This is based on evidence such as that reported in Jones-Lee and Loomes (1995) and is discussed further in the NERA/CASPAR (1998) report. (See glossary for abbreviations).

Adjusting the baseline VPF

4.40 Our approach has so far yielded an air pollution baseline VPF applicable to an average person (average age, health, wealth, etc.). Multipliers for other factors, such as advanced age, poor health and short life expectancy, might then be applied to this base.

4.41 This, however, raises a key ethical decision that has to be made in developing a valuation method: that is, whether or not to make any further adjustments at all[9]. We have stressed that our aim is to value the potential benefits of reducing air pollution from the perspective of the beneficiaries - in other words we try to predict the values that those affected would themselves use, if they were making the relevant choices and trade-offs, to reflect their strengths of preferences. Our judgement is that people's preferences for reducing the risks that could shorten their life will be stronger if they would have the chance of living, say, an extra year in good health than if they only stood a chance of living an extra year in poor health. In other words, we believe that the factors we have highlighted in Table 4.1 will affect the value people put on reducing risks, and that it is important to reflect this.

4.42 There is also the issue of how much of society's resources to devote to reducing air pollution risks, and how much to devote to other uses, including health care and other public services. In making such decisions, there are arguments for using the values that those affected would use. That said, there are other ways of approaching such decisions, and we have discussed these in Chapter 3. However, we take the view that our estimates of appropriate adjustments, using the methods we describe, do have a role to play. But we acknowledge that some may take a different stance, and that they might, for example, view our unadjusted baseline figure as more appropriate. In a sense, therefore, the unadjusted figure might be regarded as an upper bound.

The impact of age, life expectancy and attitudes to risk

4.43 There are strong theoretical and empirical grounds for believing that, in any context, the VPF can be expected to decline with age, at least after middle age[10]. Several empirical studies have produced evidence of a significant inverse relationship between the VPF and age, beyond middle years. Perhaps the most marked example is the pronounced inverted-U life-cycle for the roads VPF which emerged from the data generated by a nationally representative sample survey employing the contingent valuation (CV) approach carried out in 1982 and reported in Jones-Lee (1989). A later roads CV study reported in Jones-Lee *et al* (1993) suggests that the inverted-U life-cycle may be very much flatter than that which emerged from the earlier study. However, the results of the more recent HSE/DETR/Home Office/Treasury project roads VPF main study - completed in mid-December 1997 - substantially reinforce the findings of the 1982 study, pointing toward a roads VPF that is between 55% and 75% of its mean value for 75 year olds and between 28% and 40% for 85 year olds[11].

4.44 The simplest explanation of the inverted-U life-cycle for the VPF is that it reflects the impact of two opposing effects - namely the decrease in life-expectancy with age (which tends to depress WTP for safety) and the tendency for aversion to physical risk to increase with age (which increases WTP). The hypothesis of increasing physical risk aversion seems particularly plausible for individuals between the ages of 18 and 40, given the well known tendency for younger adults to take a somewhat more cavalier attitude towards their own safety than do those of more advanced years.

---

[9] It is worth noting that two recent studies have omitted such adjustment factors (British Lung Foundation, 1998; Ostro and Chestnut, 1998) while another study (Smith, 1998) has argued that they are relevant.

[10] In theory, within the framework of a standard economic model (discounted expected utility) the VPF will either continuously decline as age increases or will follow an inverted-U over the life cycle. The specific pattern depends upon the particular assumptions made concerning borrowing and lending opportunities and the consequent implications for the way consumption varies over the life cycle: with constant consumption the VPF is monotonically declining, whereas given constraints on borrowing and lending opportunities, as in Shephard and Zeckhauser (1982), the VPF initially rises as earnings (and hence ability to pay) increase, peaks at middle age and then declines.

[11] These figures are standardised for (i.e., do not reflect differences in) income. The ranges reflect the difference between the results of a multiple regression analysis that includes only a quadratic age variable (which produces the lower figures) and one that includes both a linear and quadratic age variable (which produces the higher figures).

4.45 In the light of these data, taking the mid-points of the ranges referred to above (paragraph 4.43), adjustment factors - to be applied to the baseline air pollution VPF of £2 million for the general population - for the combined effects of age and increasing physical risk aversion might plausibly be taken to be:

Age   Multiplier

65 -   100% of the baseline VPF
70 -   80% of the baseline VPF
75 -   65% of the baseline VPF
80 -   50% of the baseline VPF
85 -   35% of the baseline VPF

The epidemiological studies only give information on people over 65 in general. We therefore suggest using an adjustment factor weighted to take account of the average distribution of ages within the over 65 age group in the general population. This factor would be 70%. Thus, the adjusted VPF would in this case be £1.4million.

Figure 4.1   Figure 4.1 shows how the VPF is estimated to decline with age (see text above for basis of figures). It also shows how the decline parallels the decrease in remaining life expectancy.

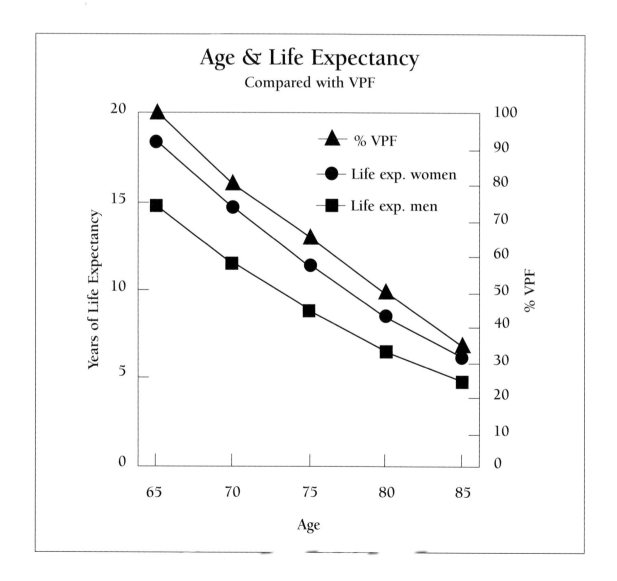

The impact of impaired health state

(i) *Life expectancy below average*

4.46 The figures in paragraphs 4.43 to 4.45 refer to results derived from the general population, which will enjoy average life expectancy. Figure 4.1 in fact shows average life expectancy at different ages - for example, at age 75 this is around 10 years (9 for men and 11 for women) and at 85 it is 5 to 6 years (for men and women, respectively). However, as previously noted (see paragraph 2.12) those who are at risk of dying from air pollution could have a very much shorter life expectancy than this - for reasons other than air pollution. Their life expectancy is not known with any certainty but is thought to be between one month or less and one year. Thus, we need to consider whether a further adjustment is needed for reduced life expectancy and, if so, what this should be.

4.47 We believe that, while life expectancy will not be the only factor affecting WTP, its effect will be positive - in other words, other things being equal, increasing life expectancy would increase WTP, and reducing life expectancy would reduce WTP. There are theoretical reasons for this: basically, that WTP is expected to be a function of, *inter alia*, aggregate future utility and the more years of life expectancy remaining, the bigger that aggregate is likely to be. This presumption is not contradicted by the data - as shown in Figure 4.1, WTP drops with age in the older age groups and reductions in life expectancy are thought likely to play a significant part in this. The more difficult question is to what *extent* WTP falls with reducing life expectancy. There is no empirical evidence on WTP by people with severely reduced life expectancy. There are several theoretical models of WTP which would predict falling WTP with reducing life expectancy but such predictions have not been tested directly. Moreover, there is debate about other features of the models, some of which would produce effects that could reinforce the predicted effect of life expectancy, and others of which might have contrary effects[12]. We did not, therefore, select any particular theoretical model but took a more general view which reflected the uncertainties. Our view is that it is unclear by how much WTP would be reduced but, on theoretical grounds (Shepherd and Zeckhauser, 1982) and given the way WTP is observed to fall with age, that it is unlikely to drop more than in proportion to life expectancy.

4.48 In the previous section, we suggested an adjustment to 70% of baseline for over-65s. Average life expectancy for this group is about 12 years (based on life expectancies for people aged 65 to 90 - see Annual Abstract of Statistics, Office for National Statistics, 1998). Thus, for over-65s with a life expectancy as little as one year, WTP would be expected, as a minimum, to be one twelfth of this (i.e., giving a factor around 6% of the baseline). For 1 month, a further division by 12 would be needed (i.e., giving a factor around 0.5% of the baseline). A factor of 6% of the baseline VPF would give a one-year figure of £120,000[13], while a one-month figure would be £10,000. These would be our minimum estimates. The maximum would be the figure of £1.4 million (see paragraph 4.45), reflecting advanced age, but not reduced life expectancy.

---

[12] For example, life expectancy as short as one year probably has the corollary of a high initial, or background, level of mortality risk. We try to assess the implications of this for the VPF - see paragraphs 4.52 to 4.54 and Annex 4A.

[13] By way of comparison, a figure estimated by Johannesson and Johansson (1997b) in Sweden was only (US) $1,500 (i.e., about £1,000) for one extra year of life expectancy (at age 86) for people who expected their quality of life during that extra year to rate at 0.45 (see paragraph 4.78). For a higher quality of life rating around our higher rating of 0.76, their figure would have been about double. Thus, our approach might seem to produce substantially higher values. However, (as noted in paragraph 4.78) the average age of respondents in this Swedish study was 42, so that the extra life-year was, on average, 44 years away. Discounting our figure at a rate of about 11% over 44 years would produce a figure similar to the Swedish result. (Discounting is explained further at paragraphs 4.64-4.65, and see the glossary.) Johannesson and Johansson inferred, from the way their results varied with the age of the respondent, a discount rate of about 1%. However, their estimation of the discount rate may have conflated a number of factors to do with how responses would change with age - e.g., factors which we discuss in this Chapter - and asking about a life-year 44 years into the future could be expected to produce responses that would be less reliable than when asking about more immediate effects.

(ii) *Quality of life*

4.49 The factors discussed in the previous section implicitly assume that the individual concerned is in an "average" state of health for someone in their age group. However, it would appear that people who die from air pollution episodes are, prior to their death, typically in a state of severely impaired health (see paragraphs 4.15 to 4.20). We therefore need to consider whether it is appropriate to apply adjustment factors that take account of the fact that health-related quality of life may be substantially reduced.

4.50 It is now well established, both in the health economics and VPF literature (e.g., Desvousges *et al*, 1996; Dolan *et al*, 1996; Jones-Lee *et al*, 1993), that there are some states of health impairment that, at least *ex ante*, are viewed by many people as being as bad as (or even worse than) death. Such states typically involve severe constraints on mobility, together with pain, discomfort, distress, and need for help with feeding, bathing, dressing and other daily activities. Willingness to pay to prolong such states might be assumed to be low. At the other end of the spectrum, it is possible to conceive of some health conditions which, though severely curtailing life expectancy, nonetheless permit a reasonable quality of life during such years or months of life as remain. In such cases it seems hard to conclude that WTP would differ, on account of quality of life, from that of the general population. We suggest that further quality of life assessment may play a useful role here.

4.51 It therefore seems highly likely that people would be willing to pay more to reduce risks if they stood a chance of, say, an extra year in good health, rather than just a chance of an extra year in bad health. The study by Johannesson and Johansson (1997b) provides some support for this assumption (see footnote 13 and paragraph 4.78). Their study found a positive and significant correlation between WTP and the quality of life that people expected to enjoy during an extra year of life expectancy. However, there are some difficulties with this study (paragraph 4.78). Given the limited amount of direct evidence, we have examined the effects of assuming that WTP is proportional to an index of health-related quality of life. As discussed at paragraph 4.16 above, our best estimate of the appropriate rating is 0.4 (± 1 SD 0.2 - 0.7) for respiratory deaths. We cannot say what factor should be used for cardiovascular deaths, so instead, for illustration, we show the implications of a quality of life rating relating to the general elderly population (see paragraph 4.20) - that for people aged 65 or over being 0.76. For the latter, given health that is average for their age, we suggest no further adjustment to the VPF. For the former, a reduction proportional to their reduced health status would be 0.4/0.76 - i.e., a factor of 53% ($\pm$ 1 SD: 26% to 92%). When applied to the figures derived above (see paragraph 4.48), figures reflecting both short life expectancy and reduced health status would be:

£63,000 (£32,000 to £110,000) for one year,

£5,300 (£2,600 to £9,200) for one month.

Thus, the range of fully-adjusted figures would span from £2,600 to £110,000 (while the figure adjusted only for age would remain at £1.4 million).

The impact of level of exposure to risk

4.52 Table 4.1 suggested that people may be keener (i.e., willing to pay more) to reduce a high risk of death by a given amount than a low risk of death by the same amount. In the groups most at risk from air pollution the background risk is higher than for the general population - for reasons other than air pollution. Some comparatively simple calculations - see Annex 4A - show that unless the background risk is very high - i.e., 50:50 or so - then adjustments to baseline VPFs for background risk will be trivial by comparison with, say, adjustments for age or health status. Annex 4A provides an analysis which could be applied where those facing a significant air pollution mortality risk also faced a very high background level of risk.

4.53 As previously noted, however, premature deaths from air pollution could involve an average loss of life expectancy of between one month or less and one year (see paragraph 2.12). People with an average life expectancy this short might well have a background risk of surviving the current year of less than 50:50, and so the effect of background risk needs to be considered. Predicting the effect of such a high background risk does, though, depend on the model from which the predictions are derived. The effect of background risk is very important in some models of WTP - in particular, in models in which there are no annuities and people are assumed to have no bequest motive, it can be shown that WTP for a reduction in risk is not only proportional to the discounted number of life years expected to remain to them but also inversely proportional to the baseline probability of surviving the current period. This probability, as noted above, is very low in the case of those with very short life expectancy. This dramatically attenuates the effect of reduced life expectancy on the VPF. (Of course another corollary of short life expectancy is probably very poor health-related quality of life, which is likely to reduce the VPF - see paragraphs 4.49 to 4.51.) Other models of WTP, on the other hand, assume that most people attach value to making bequests to those who would inherit from them if they were to die (and are not just interested in their own personal consumption during their own lifetime). In such models the level of backgound risk has very much less effect, as people are assumed to be interested in more than just their own longevity. Given that two thirds of UK households have some form of life insurance, this assumption may not be unreasonable. The prevalence of a bequest motive might be resolved by empirical research but, as this has not yet been undertaken, the bequest motive is unproven. Therefore, uncertainty remains as to the significance of background risk levels for the VPF[14]. This again serves to illustrate the sensitivity of the calculations to the impaired health state adjustment factors, and that the assumptions underlying adjustments to the baseline VPF are somewhat precarious.

4.54 That said, it is not currently possible to identify very precisely the sub-groups who are susceptible to air pollution-induced mortality or to quantify the levels of risk they face, even though it is clear that the background risks they face are likely to be higher than average. If this information becomes available, consideration may need to be given to adjustments for high background levels of risk.

The impact of wealth/income/socioeconomic status

4.55 Higher wealth or income is expected to lead to higher WTP. However, we have very serious doubts about the political acceptability of making adjustments to reflect this. In particular, we know of no other context in the UK in which such adjustments are made. The proposal that adjustments of this kind might be appropriate in the case of international safety externalities - some of which result from air pollution - has met with widespread disapproval and distaste. In the light of this, we took the view that a standard VPF should apply regardless of the income of those most affected.

4.56 There is, though, theoretical and empirical evidence on this issue which others might want to draw on. Economic theory predicts that, under a wide range of circumstances, the elasticity of the WTP-based VPF with respect to wealth or income will be greater than or equal to one - that is a 10% increase in an individual's wealth or income will raise their WTP by at least 10% (see, for example, Jones-Lee (1989) section 2.1, or the NERA/CASPAR report (1998) section 4.3).

4.57 However, as noted in the NERA/CASPAR report, while empirical studies have tended to confirm the existence of a positive and significant relationship between WTP and wealth or income, the implied income elasticities have typically been less than one and in the 0.3 to 0.6 range - so that, e.g., a 10% increase in income would increase WTP by only 3% to 6%. Again, the NERA/CASPAR report points out that these empirical estimates are based upon cross-sectional data - i.e., comparing different individuals - rather than the ideal data, which would assess the effects of variations in an individual's wealth or income, over time. While the available data are not ideal, therefore, for testing the theoretical predictions, they are

---

[14] See Jones-Lee, 1989, section 2.1.1 for a more rigorous treatment of this issue. The UK welfare reform Green Paper "New Ambitions for Our Country: A New Contract for Welfare" (Cm 3805, 1998) stated that in 1995 two thirds of households had some life insurance (chapter 1, paragraph 32), which might suggest that such households do attach some value to bequests.

nevertheless relevant to the issue. To the extent that our main concern is to arrive at adjustment factors applicable to different individuals or groups, estimates derived from cross section data are more directly relevant. Were it felt to be appropriate to make adjustments to a pollution baseline VPF for variations in wealth or income, then adjustments on the basis of an elasticity of approximately 0.5 would seem to be warranted. So, for example, someone with wealth or income twice as high as the average would be predicted to be willing to pay 50% more than the average.

4.58 Turning to the relationship between the VPF and socio-economic group, while earlier studies have tended to pick up social class effects through variations in income, a recent HSE/DETR/Home Office/Treasury-funded project has somewhat surprisingly yielded highly significant social class coefficients but *insignificant* income coefficients, with the VPF for social class III being 2.76 times that for social class IV/V while that for social class I/II is 4.40 times that for social class IV/V.

The impact of futurity and type of effect

4.59 It would appear that acute effects operate without any significant delay - probably a few days at most - following the precipitating pollution episode and impact principally on the elderly and those in already poor health. However, chronic and latent effects, if fatal, result in death some time after the onset of exposure to the pollutant concerned. It is unclear what groups are affected but it might not necessarily be limited to those already in poor health. In the case of chronic effects, death is thought to be typically preceded by a possibly protracted period of deteriorating health, while latent effects are thought typically to have no immediately perceptible impact, but eventually to lead to death from cancer after a lag of possibly many years.

4.60 In paragraphs 4.37 - 4.38, we derived a range for the air pollution baseline VPF of around £2m - £2.5m. The upper end of this range - i.e., £2.5m - would, we believe, probably reflect those risks that people dread most. We feel such a value would be appropriate in the circumstances of latent risks given that their effects, if fatal, are thought typically to entail death from cancer which would be expected to increase WTP (see Jones-Lee, 1989, and Savage, 1993). (No account has been taken of the air pollution-induced period of ill health that is typically associated with chronic effects - but morbidity impacts are dealt with separately in Chapter 6.)

4.61 Unfortunately we lack the epidemiological data about the groups at risk in the UK to be able to arrive at values of the chronic and latent risks. Their omission is a major gap, but it will be noted that the COMEAP quantification report, on whose evidence we have relied, did not feel that the evidence was sufficient to allow quantification of the consequences of the chronic risks for life and health in the UK. The COMEAP report did not consider pollutants with latent risks.

4.62 That said, though, we do regard our *method* of valuation as a useful interim way of making some progress in the absence of direct empirical evidence of WTP for risk reduction in the air pollution context. Further, we believe that were the epidemiological evidence available then our method could be used as a starting point for estimating WTP values for the chronic and latent risks. The factors that we have listed in Table 4.1 are, in principle, just as relevant to chronic and latent risks as to acute risks, and the methods we have used to allow for their effect are just as applicable as in the case of acute risks. However, we have had insufficient time to consider fully the application of our methods to the valuation of chronic or latent mortality risks. We suggest that for chronic mortality risks the baseline air pollution VPF should still be about £2m, unless the chronic risks included cancer mortality risks where we suggest a figure of about £2.5m[15] to take account of people's dread of cancer mortality over and above other mortality risks (and this last baseline VPF would apply to latent mortality risks - see paragraph 4.60). Application of the other adjustment factors would, in practice, be more difficult given there is even less information available than for acute mortality risks. However, it is plausible that those affected might be more like the population as a whole, and thus, there would not be the same need for downward adjustment as there was with the acute effects.

---

[15] Professor Ayres wished to record his unease about suggesting adoption of this figure without better empirical evidence.

4.63 Another option we briefly considered was valuing years of life lost. It is easier to express results of chronic mortality studies in terms of years of life lost. We reject the simplistic version of this (VPF divided by years lost) but did not have time to explore more sophisticated versions taking into account more factors than number of years alone. (We know, for example, that WTP depends on age.) Further thought would need to be given to this issue. In the meantime, we feel that our procedure for adjusting a VPF might reasonably be applied, with similar caveats, to the case of chronic mortality risks.

4.64 This leaves the question of how to deal with the time lag that will usually occur between the first exposure to the pollution that results in chronic and latent effects and the eventual death from such effects. If the prevention or attenuation of air pollution is to be valued at the time at which the prevention or attenuation is effected, then the problem is to determine the present value of the postponement of a fatality which would otherwise occur at some (possibly distant) future date. The key questions are, therefore, (a) what future value to associate with the prevention of a fatality a number of years away, and (b) what discount factor to apply to this future value. (Discounting reflects the fact that a benefit some time in the future is less valuable to people than an immediate benefit; see also glossary and footnote 16 below.)

4.65 Economic analysis (see Broome and Ulph, 1991; Jones-Lee and Loomes, 1995) suggests the appropriate procedure is to value the prevention of a future fatality on the basis of precisely the same VPF as would apply to the prevention of a current fatality in the same age group, and then to discount the future benefit only at the pure time preference rate for utility.[16]

4.66 Allowance should, of course, be made for the fact that the victim would be older at the end of the time lag and, therefore, the appropriate VPF should reflect the age at which the fatality would be prevented, as described in paragraph 4.45. For example, after a lag of 10 years, it would be appropriate to use the VPF of a 75-year-old rather than a 65-year-old and, after 20 years, the VPF of an 85-year-old rather than a 65-year-old, etc.

4.67 The Treasury's "Green Book" suggests that a plausible long term value of the pure time preference rate for utility would be around 1% per year (see HM Treasury, 1997, Appendix to Annex G[17]. Thus, the discount factor to be applied to the VPF for the prevention of a fatality ten years hence would be about 0.9, while that for the prevention of a fatality twenty years from the present would be about 0.8. While we are strongly of the view that the pure time preference rate for utility is the appropriate rate to use, we recognise that others may wish to see how rates other than this utility rate would affect the result, or they may have other estimates of what this utility rate is. Thus, at a 6% rate the factors would be about 0.6 and 0.3 for ten and twenty years hence, respectively (6% per year being the discount rate applied in the UK public sector - see HMT, 1997, Annex G).

---

[16] The rate at which people are willing to trade off present for future benefits is termed their *time preference rate*. A positive time preference rate means that people discount future benefits (relative to present benefits). Discount rates used by the public sector to reflect time preference are generally made up of two elements. The pure utility element reflects the natural tendency to want to have benefits sooner so as to try to avoid the uncertainty implicit in all future events. The other element of the discount rate, which reflects diminishing marginal utility of wealth at higher levels of wealth - as GDP per head rises over time - can in this case be ignored since it would simply cancel out an associated increase in the VPF at higher levels of wealth - as noted in paragraphs 4.56 and 4.57. The reason for the cancellation is essentially that the denominator in the expression for the VPF is itself the marginal utility of wealth.

[17] We note that the previous edition of the Treasury's "Green Book" put the pure time preference rate for utility at 1.5% per year (HMT, 1991). We also note that the Treasury's preferred method of allowing for futurity would involve discounting the VPF at the UK public sector discount rate of 6% per year, but then raising the value of the VPF at 3% per year - to reflect growth in income per head of 2% per year and a factor of 1.5 for the elasticity of the marginal utility of consumption (although we also note that the 1991 edition of the "Green Book" suggested growth of income per head of 2.5% per year).

Discussion

Life years lost

4.68 One way of expressing the benefit to the nation's health from reducing air pollution-induced mortality risks is to calculate how many life-years would be gained. We have estimated the current annual toll in GB from deaths from respiratory disease in terms of life years lost, as follows:

between 1,040 and 12,500 life years lost due to ozone-induced premature mortality;

between 670 and 8,100 life years lost due to PM_{10}-induced premature mortality;

between 290 and 3,500 life years lost due to sulphur dioxide-induced premature mortality.

4.69 By way of comparison, in England and Wales in 1992 there were approximately 447,000 life years lost from all deaths from respiratory disease, and approximately 138,000 life years lost from deaths from motor vehicle traffic accidents.

4.70 We have also assessed how the life years lost due to pollution-induced mortality would rate in terms of quality-adjusted life years, as quality-adjusted life years are now increasingly used as a measure of health impact. Our estimates of the annual losses of quality-adjusted life years in GB from deaths from respiratory disease are as follows:

between 210 and 8,750 quality-adjusted life years lost due to ozone-induced premature mortality;

between 135 and 5,700 quality-adjusted life years lost due to PM_{10}-induced premature mortality;

between 60 and 2,500 quality-adjusted life years lost due to sulphur dioxide-induced premature mortality.

Cost-effectiveness analysis (CEA) and cost-benefit analysis (CBA) compared

4.71 As discussed in Chapter 3, the measurement of quality-adjusted life years, as currently practised, does not make allowance for the effects of context. Instead it is commonly argued, normatively, that such factors ought not to be included, on egalitarian grounds. This means that, in practice, policy-making based on CEA using quality-adjusted life years may turn out to have different implications from policy-making based on CBA. This is because CBA would normally measure benefits taking account of people's expressed preferences. And, if people's preferences were affected by contextual issues, CBA would usually allow for this. For example, if people tended to view some sorts of risks, or risks at different ages, as worse than others - even where the aggregate number of quality-adjusted life years at stake were the same - CBA would try to reflect this view.

4.72 On the one hand, therefore, health care policy-making (which tends to use CEA) may take its objective as maximising, given available resources, the number of quality-adjusted life years lived by the population. On the other hand, environmental policy-making (which tends to use CBA) may take its objective as maximising, given available resources, aggregate benefits based on the population's preferences.

Can the quality-adjusted life years approach and WTP approach be combined?

4.73 When considering appropriate approaches to reflecting the importance of the benefits in Chapter 3, we explored whether the quality-adjusted life years approach and the WTP approach could be combined (paragraph 3.52). We now return to this in a little more detail. Quality-adjusted life years were intended to be a measure of effectiveness and are, therefore, pre-eminantly suitable for use in cost-effectiveness analysis. It might, however, be possible to use quality-adjusted life years in cost benefit analysis if a monetary value (or set of values) could be put on a quality-adjusted life year. This in turn raises the question of how to obtain such a set of values.

4.74 One approach is described in *Policy Appraisal and Health* (Department of Health, 1995; Appendix 6). This approach works backwards from the value of reducing mortality risks. The key step is to move from the valuation of life-saving to the valuation of one life-year. The assumption that is used to take this step is that a life is equivalent to a stream of life-years. So, for example, the value of saving one life-year would be put at 1/40 of the VPF with a life expectancy of 40 years (this example actually assumes a 0% discount rate, whereas using a 1% discount rate the value of a life-year would be put at 1/33 of a VPF with a 40 year life expectancy). Quality ratings can also be applied to each year to derive a value of a quality-adjusted life year rather than of a life year. *Policy Appraisal and Health* describes this technique and gives examples of the results obtained - e.g., based on the value of life-saving used by the UK government for road safety, it derives a value of a quality-adjusted life year of £23,500 (1994 prices) - which would be about £27,000 at 1996 prices. We note that this is very substantially less than the value which we have derived for a life year of £120,000 - see paragraph 4.48 - which in fact related to quality of life rated at 0.76 (the average for the elderly population), and so would be equivalent to about £160,000 (£120,000 x 1/0.76) for a quality-adjusted life year in full health.

4.75 The simple annuitisation approach used in the above example rests on the premise that "a QALY is a QALY is a QALY". It does not, therefore, allow for variations in people's preferences - for example, if people attach greater importance to life and health at particular ages, or to avoiding particular risks or causes of ill-health or death, or other reasons for variations due to factors listed in Table 4.1. Evidence is emerging, however, that people attach different values to quality-adjusted life years at different ages (e.g., Wright, 1986; Cropper *et al*, 1994; and Johannesson and Johansson, 1997a) and that some types of risks are viewed as worse than others (despite equal numbers of quality-adjusted life years being at stake). For example, in Johannesson and Johansson's (1997a) study in Sweden, people seemed to feel that an extra quality-adjusted life year for a 30 year old was about 10 times as valuable as an extra quality-adjusted life year for a 70 year old (and this view was not significantly affected by the age of the respondent).

4.76 The NERA/CASPAR report (1998) further argues that the annuitisation of a predetermined VPF into a time-stream of constant values of life-years is essentially arbitrary, since people's WTP appears not to be the result of individual thought processes involving the summing of such a constant time-stream. The evidence, overwhelmingly, is that WTP depends upon a great deal more than life span. The report suggests that particular factors to which people attach great importance include:

- emotional and personal costs to those who would be bereaved - these might be expected to peak in middle age, when there are most likely to be young or old dependents;

- the will to live, and an associated belief in the sanctity of life and the right to protection by society from certain hazards;

- concern about not achieving certain aspirations, typically over the next few months or years, but sometimes extending for longer periods - such as seeing one's children or grandchildren grow up.

The value attached to these factors cannot be expected to be simply proportional to remaining life expectancy.

4.77 For the above reasons, valuing quality-adjusted life years by annuitising a VPF has significant drawbacks. We did not, therefore, pursue this particular approach (although we have, of course, taken quality of life and life-expectancy into account in deriving our predicted WTP values). It may, however, be worth investigating in the future. A more direct approach to assessing the value that people attach to adding quality-adjusted life years to their lives, taking more account of the particular circumstances, might overcome some of the drawbacks.

4.78 One such study has been carried out by Johannesson and Johansson (1997b) in Sweden in 1995. This asked about people's willingness to buy a health intervention that would extend their own life expectancy at age 75 from 10 years to 11 years, assuming they lived to 75. The maximum that the average respondent was prepared to pay was about (US) $1,500 for this increase (1995 prices). However, the respondents tended to underestimate expected quality of life during the extra year - their expectation was that it would rate at about 0.45 (on a 0-1 scale), whereas survey findings suggested that it would on average be about 0.75. Willingness to buy the measure increased by about 30% when expected quality of life during the extra year was 0.10 greater. WTP increased by only around 0.6% for each year of age above the average age, which was 42 (although interpreting this cross-sectional variation as a time preference rate may not be entirely robust - see footnote 13). This study seems to us to provide a better approach to the value that people attach to gaining quality-adjusted life years than that discussed earlier - although it is unfortunate that for the average respondent in this study the life-year extension would have been 44 years away, which would have made it difficult for the respondent to give an accurate response.

Summary of our mortality risk valuation method

4.79 We have attempted to delineate a baseline VPF for air pollution mortality risks, to which adjustment factors can then be applied to reflect the various reasons why individual WTP for reductions in such risks would vary. Given the paucity of existing empirical evidence, it has been necessary to rely to a substantial degree on judgement and speculation. These judgements have taken account of such empirical and theoretical evidence as is available, but our recommendations should be viewed as being merely tentative guesses at how things might turn out, as and when the required empirical work has been carried out. Further empirical work is, we feel, urgently needed.

4.80 In summary, our recommendations concerning the pollution baseline VPF and the various adjustment factors are as follows:

(i) *Pollution baseline VPF*

The pollution baseline VPF should be set in a range from £2.0m to £2.5m, with a figure towards the bottom end of this range for acute and chronic effects, and towards the top end of the range for latent effects, given that the latter, if fatal, typically result in death from cancer. This baseline figure then needs to be adjusted according to the characteristics of those affected in order to predict what their WTP might be were they to be asked. We make some suggestions below for acute risks of mortality. Less is known about the characteristics of those affected by chronic or latent risks.

(ii) *Adjustment for age*

Those affected by air pollution are thought to be predominantly over 65. We have evidence to suggest that the WTP of those over 65 might be about 70% of the mean value for the population. Applying this to a baseline figure of £2m would give a value of: £2m x 70% = £1.4m.

(iii) *Adjustment for impaired health state*

Reduced life expectancy: Those affected by air pollution have a life-expectancy that is significantly lower than the average for this age group (1 year or less rather than 12 years). There are no direct studies of the WTP of people with a life-expectancy reduced below average but we believe that this will reduce the WTP of those affected. There are theoretical reasons for this and it is compatible with the fact that WTP drops with age in the older age groups. However, it is unclear to what extent WTP might be expected to drop. From theory and the relationship of WTP with age, we do not consider WTP would drop more than in direct proportion to life-expectancy.

An upper bound would be the age-adjusted figure with no downward adjustment: £1.4m.

A lower bound figure would be derived from scaling the age-adjusted figure in proportion to the reduced life expectancy:

1 year loss in life expectancy: £1.4m x 1/12 = £120,000

1 month loss in life expectancy: £1.4m x 1/12 x 1/12 = £10,000

Reduced quality of life: According to the epidemiological evidence and informed medical opinion, those who are most at risk of air pollution-induced premature death from respiratory disease are older people who already have substantially reduced quality of life as a result of health impairment due to causes other than air pollution. These people will have a lower quality of life than the average for the elderly population (0.76), and will typically be like patients with severe COPD, who have rated their quality of life at 0.4 (± 1 SD 0.2 to 0.7). Again, we would expect this to depress WTP but it is unclear by how much. For lack of good evidence, we propose that it might drop in proportion to changes in quality of life. We do not know whether WTP would drop more than in proportion to changes in quality of life, so this may not be a lower bound.

An upper bound would be the age-adjusted figure with no downward adjustment: £1.4m.

"Low estimates" would be derived from reducing WTP adjusted for age and reduced life expectancy in proportion to reductions in quality of life.

1 year loss in life expectancy; quality of life 0.2 to 0.7:
£120,000 x 0.2/0.76 = £32,000 to £120,000 x 0.7/0.76 = £110,000.

1 month loss in life expectancy; quality of life 0.2 to 0.7:
£10,000 x 0.2/0.76 = £2,600 to £10,000 x 0.7/0.76 = £9,200.

(iv) *Adjustment for background risk and wealth/income/socio-economic status*

Air pollution mortality risks are not evenly spread across the population. It is believed that those whose deaths are brought forward by air pollution, prior to their death face a substantially greater general mortality risk than the general population (for reasons other than air pollution). If it were possible to identify accurately the sub-groups who face the highest risk of air pollution-induced mortality and to quantify the level of background risk they face, consideration would need to be given to adjustments to the VPF for high background levels of risk although there are uncertainties over what adjustment would be appropriate. In the meantime, we do not suggest adjustment for background risk.

For variations in wealth, income or socio-economic status, both theory and empirical evidence suggest that non-trivial multipliers *would* be called for - were it felt appropriate to make adjustments for these factors. However, we have very serious doubts about the political acceptability or ethical appropriateness of making such adjustments.

Overall suggested estimate for air pollution VPF

Summarising the above suggests an estimate for an air pollution VPF below £1.4m. How far below is uncertain but it could be as low as £2,600 for a 1 month loss of life expectancy and a quality of life of 0.2. The derivation of these figures is summarised in Figure 4.2.

Figure 4.2   Summary of possible adjustments

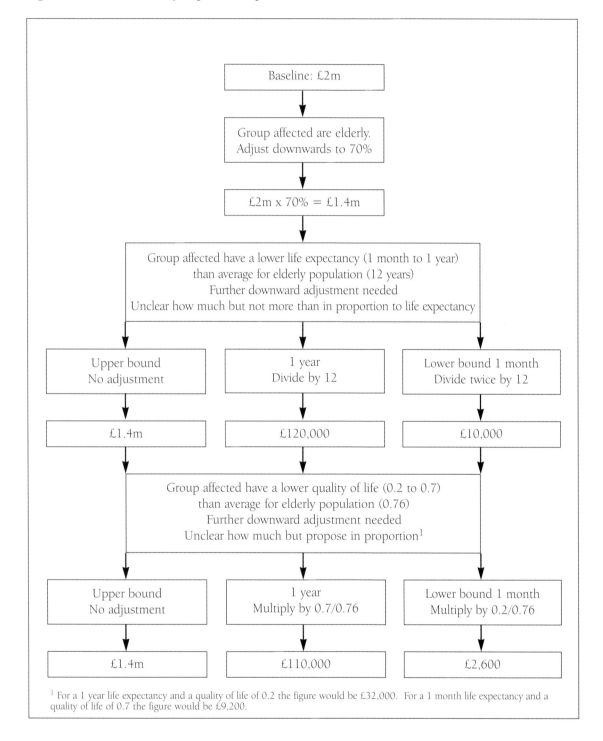

4.81 The above adjustments for life expectancy and quality of life relate to respiratory deaths. For cardiovascular deaths it is unclear what types of heart disease are affected and this can have a significant effect on what adjustment factors would be appropriate. For example, it is possible to have diseased coronary arteries with no apparent effect on quality of life until a sudden heart attack. We do not consider that we can specify particular adjustment factors but in general terms it is likely that there would be less downward adjustment for cardiovascular deaths.

4.82 Cardiovascular deaths are more common than respiratory deaths but, since air pollution has a greater effect on respiratory deaths (paragraph 2.20), it seems unlikely that the deaths brought forward by air pollution are dominated by cardiovascular deaths. We therefore suggest that figures for respiratory deaths are used for valuation of all-cause mortality but that it is noted that, if adjustments are made for the reduced life expectancy and quality of life expected in those with respiratory disease, the resultant value is likely to be an underestimate because some of the deaths will be cardiovascular deaths.

Futurity and Type of Effect

4.83 In the case of chronic and latent effects, were they to be quantified, it would be necessary to take account of the fact that those who die as a result of such effects typically do so at some considerable time after the first exposure to air pollution. As far as futurity and discounting are concerned, in the case of safety effects there are persuasive arguments for valuing future risk reduction benefits at current values, discounted *only* at the pure time preference rate for utility. Taking a value for this of 1% per annum, the following discount multipliers are implied:

Futurity of Impact	*Multiplier* @1% discount rate
5 years	95%
10 year	90%
15 years	86%
20 years	82%
25 years	78%

Gaps and limitations

4.84 We would not want to give the misleading impression that our analysis here can be considered to be a definitive statement on the valuation of air pollution-induced mortality risks. Our analysis is based on many rough and ready estimates and methods, and we reiterate the point made earlier (paragraph 4.79) that we view our findings as tentative guesses as to the appropriate valuations. In several areas our analysis is likely to be incomplete. Relying only on health effects that can currently be estimated with some confidence would, under most plausible assumptions, produce underestimates (perhaps severely so) of the total health effects and their associated value. Some areas where important evidence is lacking include:

- chronic mortality - which, if it were to prove as significant as has been suggested (2.25), might dominate any future valuation exercise;

- latent mortality - about which little seems known;

- acute deaths in sub-groups other than the very frail - e.g., if the epidemiology were to indicate the possibility of air pollution reductions reversing, say, cardiovascular conditions in the young - although the numbers may be much smaller, the valuations might be much larger;

- direct evidence of valuations of reducing mortality risks by people in various different states of health;

- direct evidence of valuations of reducing air pollution mortality risks compared with other types of mortality risks.

We recommend these as areas for research.

References

Beattie, J., Chilton, S., Cookson, R., Covey, J., Hopkins, L., Jones-Lee, M., Loomes, G., Pidgeon, N.F., Robinson, A. and Spencer, A. (1998) *Valuing Health and Safety Controls: A Literature Review*. London: HSE Books.

British Lung Foundation (1998). *Transport and Pollution - the Health Costs*, London: British Lung Foundation.

Broome, J. and Ulph, D. (1991). *The Intergenerational Aspects of Climate Change*. Bristol: Department of Economics, University of Bristol.

Cropper, M.L., Aydede, S.K. and Portney, P.R. (1994). Preferences for life saving programs: how the public discounts time and age, *J. Risk Uncertainty* **8**: 243-65.

Dalvi, M. Q. (1988) *The Value of Life and Safety: A Search for a Consensus Estimate*. London: Department of Transport.

Department of Health. Committee on the Medical Effects of Air Pollutants (1998). *Quantification of the Effects of Air Pollution on Health in the United Kingdom*. London: The Stationery Office.

Department of Health (1995). *Policy Appraisal and Health. A Guide from the Department of Health*. London: Department of Health.

Department of Health (1998). *Health Survey for England 1996*. London: The Stationery Office.

Desvousges, W.H., Reed Johnson, F., Hudson, S.P., Gable, S.R. and Ruby, M.C. (1996) *Using Conjoint Analysis and Health-State Classifications to Estimate the Value of Health Effects*. Durham, NC: Triangle Economic Research.

Dolan, P. Gudex, C., Kind, P. and Williams, A. (1996) Valuing health states: a comparison of methods. *J. Health Econom.* **15**: 209-231.

HM Treasury (1991). *Economic Appraisal in Central Government: A Technical Guide for Government Departments*. London: HMSO.

HM Treasury (1997). *Appraisal and Evaluation in Central Government*, London, The Stationery Office.

Johannesson, M. and Johansson, P.-O. (1997a) Is the valuation of a QALY gained independent of age? Some empirical evidence. *J. Health Econom.* **16**: 589-599.

Johannesson, M. and Johansson, P.-O. (1997b) Quality of life and the WTP for an increased life expectancy at an advanced age. *J. Public Econom.* **65**: 219-228.

Jones-Lee, M.W. (1989) *The Economics of Safety and Physical Risk*. Oxford: Basil Blackwell.

Jones-Lee, M.W. (1992) Paternalistic altruism and the value of statistical life. *Econom. J.* **102**: 80-90.

Jones-Lee, M.W., Loomes, G., O'Reilly, D. and Philips, P.R. (1993) *The Value of Preventing Non-Fatal Road Injuries: Findings of a Willingness-to-Pay National Sample Survey* TRL Working Paper WPSRC2.

Jones-Lee, M.W., Loomes, G. and Philips, P.R. (1995) Valuing the Prevention of Non-Fatal Road Injuries: Contingent Valuations vs Standard Gambles. *Oxford Economic Papers* **47**: 676-695.

Jones-Lee, M.W. and Loomes, G. (1995) Discounting and Safety. *Oxford Economic Papers* **47**: 501-512.

Measurement and Valuation of Health Group (1995). *The Measurement and Valuation of Health*. York: University of York, Centre for Health Economics.

NERA/CASPAR (1998) *Valuation of Deaths from Air Pollution*. London: NERA.

Office for National Statistics (1998). *Annual Abstract of Statistics 1998*. London: The Stationery Office.

Office of Population Censuses and Surveys (1994) *Mortality Statistics, Series DH1 No.28*. London: HMSO.

Osman, L.M., Godden, D.J., Friend, J.A.R., Legge, J.S. and Douglas, J.G. (1997) Quality of life and hospital re-admission in patients with chronic obstructive pulmonary disease. *Thorax* **52**: 67-71.

Ostro, B. and Chestnut. L. (1998) Assessing the health benefits of reducing particulate matter in the United States, *Environ. Res.* **76**: 94-106.

Savage, I. (1993) An empirical investigation into the effects of psychological perceptions on the willingness-to-pay to reduce risk. *J. Risk Uncertainty* **6**: 75-90.

Shephard, D.S. and Zeckhauser, R.J. (1982) Life-Cycle Consumption and Willingness to Pay for Increased Survival. In: *The Value of Life and Safety: Proceedings of a Conference Held by the Geneva Association*. (Jones-Lee, M.W. ed). Amsterdam: North-Holland.

Smith, A.E. (1998) *Valuing Mortality Risks Associated with Air Pollution*. Washington, DC: DFI Aeronomics.

TRL (1997) *Highways Economics Note No 1*.

Viscusi, W.K., Magat, W.A. and Huber, J. (1991) Pricing Environmental Health Risks: Survey Assessments of Risk-Risk and Risk-Dollar Trade-Offs for Chronic Bronchitis. *J. Environ. Econom. Manage.* **21**: 32-51.

Viscusi, W.K. (1993) *Fatal Tradeoffs*. New York: OUP.

Wright, S. J. (1986) *Age, Sex and Health: A Summary of Findings of the York Health Evaluation Survey*. Discussion Paper 15. York: University of York, Centre for Health Economics.

ANNEX 4A
Baseline Risk

1. Table 4.1 suggested that people may be keener to reduce a high risk of death by a set amount than a low risk of death by the same amount. We have, therefore, considered whether this effect means an adjustment factor is needed when considering deaths brought forward by air pollution as compared to road accident deaths. Modelling of the expected change in WTP according to background level of risk (see box) suggests that there is no marked effect over a wide range of levels of risk, the effect only beginning to be more significant above levels of risk around 50:50.[18].

> **How WTP Varies with Background Risk**
>
> Some insights can be gained from a simple single-period standard economic model.[1] In particular, the VPF for a group of individuals each of whom faces an initial probability p_1 of death by a given cause will be $(1 - p_2)/(1 - p_1)$ times the VPF for a group of individuals each of whom faces an initial probability p_2 of death by the same cause. Furthermore, for p_1 and p_2 both small, to a good approximation we can write $(1 - p_2)/(1 - p_1) = 1+p_1-p_2$. Thus, suppose that we set p_2 equal to the annual risk of death in a road accident for the average car driver or passenger - ie 6×10^{-5}. It then follows that someone who faces an annual risk of death of 100×10^{-5} of death in a road accident would, *ceteris paribus*, have a VPF approximately 0.1% larger than the VPF for the average car driver or passenger.
>
> *1. A single-period expected utility model focuses exclusively on the risk of death in the coming period (e.g., the coming year). Thus, as far as survival is concerned, there are just two possible outcomes, namely that the individual survives the coming period or that the individual dies during the coming period. This contrasts with a multi-period model in which the possibility of death in any of several future periods (up to a maximum conceivable remaining lifespan) is explicitly accommodated).*

2. As discussed in Chapter 2 (Annex 2A, paragraph 5), the level of risk of a death brought forward by air pollution can be estimated as around 2 in 10,000 for the general population and around 10 in 10,000 for the elderly population. Following the method in the box, starting with a background risk of 10 in 10,000 rather than the risk of 0.6 in 10,000 for road accidents would only be expected to increase WTP by about 0.1%. This is trivial by comparison with, say, adjustments for age or health status. The risk of death from any cause in the elderly is about 6 in 100 compared with 1 in 100 in the general population (Office for National Statistics, 1998). This would only increase WTP by about 5%. Thus, we do not consider that adjustment for background level of risk is required in cases like this.

3. It was also noted in Chapter 2 that those who are already seriously ill with heart or lung disease are thought to be most at risk. It is not currently possible to identify the susceptible sub-groups very precisely or to quantify the levels of risk they face, but the risks are likely to be higher than those for the general elderly population. If this information is available in future, consideration may need to be given to adjustments for initial levels of risks above 50:50.

---

[18] At these extremely high levels of risk, WTP becomes much more sensitive to particular assumptions in the model

Figure 4A.1 Willingness to pay and background risk

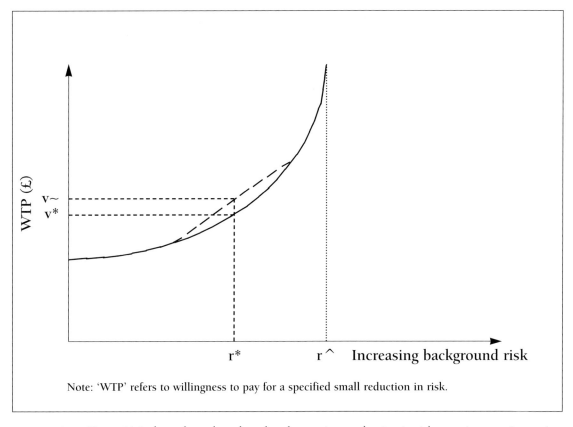

Note: 'WTP' refers to willingness to pay for a specified small reduction in risk.

4   Figure 4A.1 shows how the value placed on a given reduction in risk may rise at an increasing rate as the size of the background risk increases. (This is just a corollary of saying that people may be keener to reduce a high risk of death by a set amount than a low risk of death by the same amount.) The graph also illustrates how the average v~ of the value placed on a given reduction from a high background risk, on the one hand, and from a low background risk, on the other, may be greater than the value v* placed on the same reduction from the mid-point of those two background risks. This indicates why it is relevant to consider whether any groups are at especially high risk. However, our assessment of the current evidence suggests that we are at the "flattish" part (left hand side) of the graph in terms of the impact of air pollution risks (see paragraph 2 of this Annex). Future research may, though, reveal some especially vulnerable people - whose background risk is, say, nearer to r^ - and, if so, this might imply a need for an increase significantly more than the 0.1% or 5% adjustment derived above.

5   Some of those at very low risk in the current year may become much more susceptible in future years, if they eventually come to suffer (for other reasons - e.g., through smoking) from severe heart or lung disease. We feel it would be reasonable to assume that the minority of the current low risk group who will eventually face high risk are likely then to view the magnitude of that risk in the same way as those currently at high risk.

Chapter 5

Benefits of Reducing NHS Costs and Other Costs

Introduction

5.1 In this chapter we consider the possible benefits flowing from reduced air pollution in terms of lower costs to the NHS and to other parts of the economy. Thus, we look at the financial costs of dealing with the adverse effects of air pollution on health and how, if air pollution were reduced, such expenditures would be affected - rather than the direct effects on life expectancy and health. For the NHS there are costs associated with, for example, hospital in-patient stays, out-patient visits, general practitioners' time, and use of pharmaceuticals. Costs that fall on NHS budgets are in the end borne by taxpayers and, therefore, they are not costs which are likely to be considered in WTP estimates[1].

5.2 On top of these NHS costs, there may be some costs to patients themselves - e.g., travelling to see the doctor. There may also be other costs to society - e.g., businesses might lose the time of workers if they were incapacitated due to air pollution. Ultimately, this would mean that less output would be produced.

5.3 One issue that needs to be addressed is whether the inclusion of financial costs would indirectly double-count the measures of benefit based on people's willingness to pay (WTP), which we have already included in Chapter 4 on mortality risk reductions and Chapter 6 on morbidity risk reductions. We do expect people's WTP responses to be affected by the level of income or wealth they expect to enjoy - see paragraphs 4.53 and 4.54. However, we do not expect short periods of time off work to make such an impact on expected wealth that it would affect most people's WTP responses (and many people are anyway protected from the financial consequences through the receipt of sick-pay, so that this cost is borne by others - taxpayers or employers). We take a similar view of out-of-pocket expenditures on, say, travelling to see the doctor. We therefore think that it is reasonable to add up all these costs without much risk of double-counting. Most of those affected by the acute effects of air pollution are likely to be retired. However, were we able to draw on more evidence about the effects of air pollution on the chronic ill-health of people of working age - especially self-employed people - we feel that there would be a greater risk, in such a case, of double-counting the benefits of reduced sickness absence and WTP to avoid the chronic morbidity. Given that we have not counted such chronic ill-health effects at all, however, this does not pose a problem.

Effects on the NHS

5.4 The costs of respiratory diseases are a substantial burden to the NHS. For example, the publication "*Burdens of Disease*" (Department of Health, 1996) estimated that respiratory diseases account for around 6% of NHS hospital costs and around 12% of NHS primary care expenditure (including primary care pharmaceuticals).

5.5 The DH Committee on the Medical Effects of Air Pollutants (Department of Health, 1998) estimated that air pollution contributed to many thousands of NHS hospital admissions. They estimated that particulate matter could be responsible for 10,500 respiratory hospital admissions; sulphur dioxide could be responsible for 3,500 respiratory hospital admissions; and ozone could be responsible for perhaps 9,900 respiratory hospital admissions

---

[1] Some costs of NHS pharmaceuticals are of course paid through patients' prescription charges, although only a minority of the population are liable for such charges, and such charges make up only a small fraction of overall NHS costs.

5.6 Evidence is beginning to emerge concerning admissions for diseases other than respiratory diseases - for example, heart or cerebrovascular disease. However, the above Committee did not feel that at the present time this evidence was robust enough for them to be able to quantify the impact on NHS hospital admissions. To assist future studies, however, when more evidence may be available, and for sensitivity analyses, we discuss the costs of admissions for cardiovascular disease.

5.7 Extra hospital admissions for respiratory disease may well be associated with extra out-patient visits, visits to GPs, and extra consumption of pharmaceuticals (e.g., use of bronchodilators). Again, however, quantitative data about the actual numbers involved are not yet available.

NHS Unit Costs

5.8 The main evidence that we have available concerns spells in hospital for respiratory disease. We have, therefore, estimated costs of such hospital treatment. (See Table 5.1.)

Table 5.1 **Spells in NHS hospitals by patients admitted as emergency admissions, with a main diagnosis on discharge of diseases of the respiratory system, and NHS costs (at 1996/97 prices)**

NHS, England, 1994/95, Primary diagnosis ICD(9): 460-519: diseases of respiratory system	<65 years	65+ years	All Ages
Spells (emergency; (thousands)	247	160	407
Average length of stay (days)	3.9	13.6	7.7
Cost per spell: £	705	2,460	1,390
Total cost: £M	174	393	566

Table 5.1 takes data from the NHS Hospital Episode Statistics 1994/95. It shows numbers of emergency admissions for those aged under 65 and those aged 65 and over. (NB The table does not refer to the numbers of people admitted to hospital as some people are admitted several times.) There were 50% more emergency admissions among under 65 year-olds as among those 65 and over. However, when admitted, those 65 and over stayed nearly four times as long in hospital - two weeks on average, compared with only about 4 days for those under 65.

This Table also shows costs to the NHS. The cost per day in 1994/95 was put at £169 (Chartered Institute for Public Finance and Accountancy (CIPFA) Health Database), or about £181 at 1996/97 prices, for the average cost per inpatient day in a thoracic ward. The average cost per stay for an emergency respiratory inpatient was thus calculated to be about £1,400, while the average cost for someone 65 or over was nearly £2,500 (1996/97 prices). Total NHS hospital expenditure on emergency respiratory inpatients was calculated to be over £500m.

As mentioned in paragraph 2.33, the epidemiological studies mainly examine emergency hospital admissions as these are most likely to be affected by day to day changes in air pollution, so the costs of emergency admissions are most relevant. The table indicates, for example, that the NHS cost of an emergency respiratory admission for someone aged 65 years or older is nearly £2,500. The average cost of treating someone under age 65 is lower - because they tend to spend a shorter time in hospital owing to their younger age and the different spectrum of diseases affecting younger age groups. The cost of an emergency respiratory hospital admission averaged across all ages is around £1,400 (1996/97 prices). We suggest this is a reasonable unit cost to use as a first approximation. This assumes that the age distribution of those admitted to hospital as a result of air pollution is the same as for emergency respiratory patients in general. This is probably not the case as some pollutants have a greater effect on hospital admissions in the over 65s (see Annex 2A). In addition, while there are more respiratory admissions in those under 65, the under 65s make up a larger proportion of the population (around 84%). The emergency respiratory admission *rates*, per 1,000 population in the relevant age group, are over 3 times higher in the 65+ age group than in the under 65s. The combination of the greater effect of some pollutants and the higher baseline rate in the over 65s, together with the higher costs of the admissions, may mean that disaggregating by age gives a higher estimate of NHS costs per unit concentration of pollutant. We suggest this is examined in a sensitivity analysis.

5.9 We have also looked at hospital costs for heart disease. Although cardiovascular admissions were not quantified in the COMEAP report, hospital costs for these may be useful for sensitivity analyses. As it is unclear exactly what type of heart disease is affected by air pollution we have considered emergency admissions for ICD codes 410-429: ischaemic and other heart disease. Table 5.2 shows that the cost for a spell in hospital averaged for all ages is about £1,500 (1996/97 prices).

Table 5.2 **Spells in NHS Hospitals by patients admitted as emergency admissions, with a main diagnosis on discharge of heart disease, and NHS costs (at 1996/97 prices)**

NHS, England, 1994/95, Primary diagnosis ICD(9): 410-429: ischaemic & heart disease	<65 years	65+ years	All Ages
Spells (emergency; (thousands)	114	248	362
Average length of stay (days)	6.4	10.6	9.2
Cost per spell: £	1,030	1,710	1,485
Total cost: £M	117	424	538

Table 5.2 takes data from the NHS Hospital Episode Statistics 1994/95. It shows numbers of emergency admissions for those aged under 65 and those aged 65 and over. (NB The table does not refer to the numbers of people admitted to hospital as some people are admitted several times.) There were over twice as many emergency admissions among those 65 and over as among under 65 year-olds. When admitted, those 65 and over stayed nearly twice as long in hospital - nearly 11 days on average, compared with only about 6 days for those under 65.

This Table also shows costs to the NHS. The cost per day in 1994/95 was put at £151 (Chartered Institute for Public Finance and Accountancy (CIPFA) Health Database), or about £161 at 1996/97 prices, for the average cost per inpatient day in a medical ward. The average cost per stay for an emergency ischaemic heart disease inpatient was thus calculated to be about £1,500, while the average cost for someone 65 or over was about £1,700 (1996/97 prices). Total NHS hospital expenditure on emergency ischaemic heart disease inpatients was calculated to be over £500 million. (If costs per day in a cardiology ward had been used instead the costs would have been over twice as high - at £344 per day.)

In paragraph 2.14, we did not suggest a sensitivity analysis on disaggregation by age for cardiovascular admissions which are themselves only being covered as a sensitivity analysis. However, some feel for the possible variation in NHS costs can be seen from the range in costs per emergency admission from about £1,000 to about £1,700 for under and over 65s (1996/97 prices).

Other NHS costs

5.10 We have, unfortunately, very little information about the cost of other forms of health care attributable to air pollution. A study by Buckingham and others (1994) suggests that a year of primary health care asthma treatment for someone with asthma would cost around £60 (1991/92 prices; or about £73 at 1996/97 prices). It would seem appropriate to attribute perhaps some proportion of such a cost to asthma problems exacerbated by air pollution. However, it is not clear what proportion particularly since it is plausible that the more severe asthmatics are more affected by air pollution. Other health care use attributable to air pollution may include extra use by asthmatics of bronchodilators. The unit costs of bronchodilator use are very low - around £0.01 per dose (British Medical Association, 1998) - but for more severe asthmatics inhalers may cost the NHS around £25 per month (Professor Ayres, personal communication; see also footnote 1). The cost of an out-patient visit tends on average to be roughly one eighth of the cost of a day in hospital. On this basis, an out-patient visit for respiratory or cardiovascular disease might be put at around £21.

5.11 From the above unit costs, it would seem that in-patient costs are likely to be one of the most important sources of costs to the NHS for exacerbations of respiratory disease associated with air pollution. However, it should be borne in mind that low NHS expenditure for more minor effects could become significant if air pollution was frequently contributing to minor effects in a large number of people. At present, the evidence for effects of air pollution on respiratory symptoms is not clear enough to give guidance on this.

Aggregate NHS Savings from Reductions in Pollutants

5.12 We have calculated reductions in NHS emergency hospital admissions that would arise from hypothetical 1 $\mu g/m^3$ reductions in particular pollutants, and have then used these to assess the possible aggregate savings to the NHS (by multiplying the number of admissions by the cost per emergency respiratory hospital admission for all ages (£1400)). The results are collected together in the final chapter. Describing current costs incurred as potential savings does, however, assume that if, as a result of less air pollution, there were fewer emergency admissions then the NHS would be able to spend less. Clearly some NHS costs (e.g., consumables) could be reduced very quickly in response to such a reduction in admissions, but other costs that are, for example, linked to the size of hospital capacity would take much longer to reduce. There may, however, be benefits in the short term in forms other than financial savings, such as helping the reduction in waiting lists for other types of cases. So we feel that even if in the short term savings do not materialise in the form of less expenditure, our estimates can be thought of as a measure of the value of, say, the freeing up of hospital capacity for other uses.

Costs to Patients/Sufferers

5.13 Unfortunately, very little evidence is available about the costs in financial terms to patients and/or people who are affected by air pollution. We expect that the costs might include extra expenditure on, for example, additional visits to the health care system. (DETR estimate the average value of one hour of own time spent travelling instead of on chosen leisure activities at about £3.50 (1996 prices), and fares or petrol costs would be additional to this.) Again, therefore, we expect these costs to be of a lower order of magnitude than NHS in-patient costs - perhaps a few per cent of the latter.

5.14 Avertive behaviour represents potentially another category of costs (about which data are also lacking). It is quite possible that some people alter their lifestyles and incur costs - from buying cycling masks, to moving home or job - to try to avoid the effects of air pollution. The costs to which such avertive behaviour gives rise are potentially reducible through curbing air pollution. Unfortunately we are unable to estimate their magnitude. If any financial costs are reflected in WTP estimates (see paragraph 5.3) it is most likely to be these.

Other Economic Costs

5.15 The costs that fall on others could, in principle, include losses of work output. However, for deaths brought forward by air pollution, when we consider the state of health of those who are likely to be affected we find that it is thought most likely that such people are in a very serious state of ill-health already. For example, it is thought probable that their remaining life expectancy - with or without suffering the ill effects of pollution - may be only one month to one year. In view of this, we think it is unlikely that a significant proportion would be working even if they were below retirement age. The vast majority, though, are expected anyway to be above retirement age. For these reasons we think that losses of work output due to air pollution mortality are likely to be negligible. The same point applies to some extent to hospital admissions where the elderly and seriously ill will be most affected. It is possible that some of those admitted to hospital could be younger and less seriously ill (and, therefore, working). We have insufficient information to define the numbers in this category but it is probably small. Some of the more minor effects of air pollution which have not been quantified might also result in some time off work.

5.16 The conclusion regarding lost output and air pollution mortality could change, however, if we knew more about the effects of air pollution on chronic ill-health. For example, if air pollution is a cause of, or exacerbates, diseases such as chronic obstructive pulmonary disease amongst the working population, then there may be significant additional economic costs due to increased time off work.

Other Consequential Effects

5.17 Another issue concerns costs in years of life added when, owing to reductions in mortality risks, people's lives are extended. Guidance on the conduct of cost-effectiveness analysis in health care suggests that it can be appropriate to include costs to the NHS in added years of life[2] (and, some argue, the entire difference between individuals' consumption and their production (see, e.g., Johannesson and Meltzer, 1998)) although such an approach remains controversial. Such extra NHS costs due to increased life expectancy would, if included, need to be subtracted from the savings due to reduced morbidity outlined above. We show results with and without such extra NHS costs (see Table 7.2).

5.18 The main reason for considering such costs is that they will not have been included in WTP estimates, as they are borne by the state rather than by the individual. In principle other real costs borne by the state for such people ought also to be included (e.g., social services costs).

5.19 The DH Annual Report (Department of Health, 1998) suggests that NHS hospital and community health services expenditure per person for people aged 85 or over was approaching £2,500 per year. (The figure is lower for younger age groups. But the highest figure for this oldest age group would seem most appropriate if those whose lives were extended due to reductions in air pollution would only have a life expectancy of one year or less, since their state of health is likely to be most comparable with that of the population in the oldest age group. And as health is not expected to improve in any added months or years of life, we would not expect NHS use to decrease.) So, the prevention of a death caused by air pollution may increase NHS expenditure in the subsequent year by perhaps £200 to £2,500 (for one month to one year of added life expectancy). It is likely that social services costs for such people will be of a similar order of magnitude, and could thus double these estimates.

Conclusions

5.20 We conclude that the costs to the NHS of in-patient care of people hospitalised for respiratory disease represent the largest set of financial expenditures that we can currently identify which could be associated with air pollution. Costs for cardiovascular admissions could be of a similar order were they to be quantified in future. There are lower costs for minor effects but these could accumulate if larger numbers of people were affected more frequently than for hospital admissions. The effects of air pollution on respiratory symptoms have not been quantified so this is unclear.

---

[2] Markandya and Pearce (1989), for example, in calculating the social costs of smoking, state that "future savings which would result from the premature death of a smoker have to be credited to the overall medical costs".

References

British Medical Association. Royal Pharmaceutical Society of Great Britain. (1998) *British National Formulary. Number 35.* London; Wallingford, Oxon: British Medical Association, The Pharmaceutical Press.

Buckingham, K., Drummond, N., Cameron, I., Meldrum, P. and Douglas, G. (1994) Costing shared care, *Health Services Management* February 1994; 22-25.

Department of Health Committee on the Medical Effects of Air Pollutants. (1998) *Quantification of the Effects of Air Pollution on Health in the United Kingdom.* London: The Stationery Office.

Department of Health (1996). *Burdens of Disease: A Discussion Document.* London: NHS Executive.

Department of Health (1998). *Department of Health Annual Report.* London: The Stationery Office.

Johannesson, M. and Meltzer, D. (1998) Some reflections on cost-effectiveness analysis, *Health Economics*, **7**: 1-7.

Markandya, A. and Pearce D.W. (1989) The social costs of tobacco smoking. *Br. J. Addict.* **84**: 1139-50.

Office for National Statistics (1997). *Family Spending 1996-97*, London: The Stationery Office.

Chapter 6
Benefits of Less Morbidity

Introduction

6.1 While air pollution may bring deaths forward in some people, it is likely to cause many more to be ill. Ideally in our assessment of the morbidity due to air pollution we would have wanted to cover the full breadth and depth of the impacts. This would include both major morbidity, such as that which causes pain and severely limits people's ability to perform normal daily activities and might require hospitalisation, through to more minor morbidity which may have unpleasant symptoms but may only limit activities to a moderate extent. As with the effects of air pollution discussed in the previous chapters, we would also ideally have wanted to draw on evidence about the preferences of those who would stand to benefit from better air quality in order to reflect the value that they would attach to improvements. Unfortunately, owing to limited availability of data, our coverage of morbidity is much more incomplete than the ideal, and we have had to use indirect methods to try to estimate the values that the beneficiaries would attach to reduced risks of morbidity[1]. We believe, though, that our estimates provide a useful starting point.

6.2 There is what might be termed "a pyramid of morbidity" since people are subject to many minor ailments and fortunately to fewer major ones. Inevitably impacts on morbidity are much more difficult to trace and to quantify than impacts on mortality because, for example, the effect may be harder to define and more subjective and there is a large range of possible events. Unlike mortality, morbidity represents a continuum of different degrees of ill-health. A result is that the impact of pollution on morbidity, particularly for more minor symptoms, is more difficult to study and is less well understood than the impact on mortality.

6.3 This Chapter does not attempt to assess the quality of the epidemiological evidence, or the extent to which it is possible to derive quantified estimates of the effects of air pollution on morbidity in the UK. For this we rely on the COMEAP quantification report (see Chapter 2). Rather, we aim mainly to illustrate here how quantified evidence on the morbidity consequences of air pollution could be brought into economic appraisal, were the evidence reliable enough. These illustrations are not an alternative to new studies aimed at providing better and more direct evidence, which we believe are still needed, but an attempt to make good and apt use of results which already exist.

Morbidity from air pollution

6.4 Focusing on the central concern of our report - the impact of changes in air pollution on health - we note that, while there is potentially a wide range of possible effects to consider, the COMEAP quantification report identified only one firm impact on morbidity - respiratory hospital admissions (see paragraph 2.4). Drawing on the best evidence of the impact of air pollution in GB, it estimated the number of annual extra, or earlier, hospitalisations caused by certain pollutants as follows:

PM_{10} 10,500;

SO_2 3,500;

O_3 500-9,900.

These are the most reliable estimates of morbidity effects that we have.

---

[1] Professor Ayres wished to record that he was unhappy about a number of assumptions and extrapolations from data made in this chapter. His concerns focused on the use of predicted quality of life scores in lieu of any empirical data and the use of WTP estimates based on studies of the general population rather than of patients. He argued that the calculations shown in the chapter should be regarded as illustrative rather than indicative.

6.5 Nevertheless, it is recognised that only those with the most serious morbidity effects are likely to be hospitalised, and that there is likely to be a range of other, generally less serious, effects. For example, extrapolating from the *Health Survey for England 1996* (Department of Health, 1998), around 1.5 million adults in England would say that their asthma or wheezing attacks could be precipitated by traffic fumes[2]. The range of possible effects that has been described in the literature (e.g., Dockery and Pope, 1994) covers, for example, asthma attacks, breathing trouble, coughing, pain, eye irritation, exacerbations of conditions such as bronchitis, and being unable to perform work or other activities. We have no reliable data about the extent to which these are brought on by air pollution in the UK. However, we consider them for illustration, as better evidence may emerge in the future.

6.6 We see our task as being to suggest whether values can be applied to the prevention of morbidity associated with respiratory hospital admissions. We also consider the extent to which, if firmer estimates could be made of the other effects and the extent to which they are genuinely attributable to air pollution, this might have a significant impact on the aggregate benefit of reduced harm from air pollution. In this way we hope to establish a framework which could be used as and when further evidence accumulates.

Quantification methods

6.7 Morbidity involves a reduction in health-related quality of life. In order to gauge the extent of morbidity, and the significance of potential reductions, morbidity in its many forms needs to be described and assessed in ways that are consistent and comparable. One of the problems in this area is that there are many different, non-comparable ways of describing the effects on health.

6.8 There are standard descriptions often used in the air pollution context that range from being specific as to the particular ill-health condition, to being somewhat ambiguous:

- medical descriptions (typically for major/serious ill-health) - see, for example, Tables 5.1 and 5.2 which report health conditions defined using ICD codes;

- vaguer descriptions (typically for minor health states/events) - e.g., "restricted activity days".

Even though ICD coding may be precise as to the type of disease, it does not tell us how bad the patient actually feels, or whether one condition is worse than another. "Restricted activity days" are an example of a measure that tries to focus more closely on how the sufferer is actually affected, but which may cover such a wide range of possibilities - from being bedridden to relatively minor restrictions - that it is open to widely differing interpretations.

6.9 Health care use may be considered as an indicator of morbidity - e.g., visits to the doctor. However, such measures are difficult to interpret as measures of morbidity, and provision and use of health care is affected by more than just morbidity.

---

[2] The *Health Survey for England 1996* reported that 21% of adults and 18% of children had had asthma or wheezing attacks in the 12 months before the interview. Of adults with a history of asthma or wheezing attacks, with the most recent attack not more than 5 years before the interview, 21% reported that traffic fumes could precipitate their attacks. While it has not been demonstrated scientifically that their views about the causes of asthma attacks are correct, this finding does give an indication both of what the public believes about the consequences of air pollution and the scale of the perceived problem.

6.10 There are also generic systems for describing morbidity, such as through the use of life expectancy weighted for health-related quality of life (as discussed in Chapter 3 - see especially paragraph 3.17 and Annex 3A). For example, the EQ5D has five standard dimensions of health:

- mobility;
- self-care;
- usual activities;
- pain/discomfort;
- anxiety/depression.

The "Quality of Wellbeing" (QWB) is another such measure of health-related quality of life.

6.11 As previously discussed, advantages of generic systems are that all health states are described using the same standard descriptors. This means that health states can be compared, differences are highlighted, and evidence can be more easily transferred between contexts.

6.12 Further, if the range of health states has been ranked or scaled, they can be compared in terms of seriousness or undesirability. Health utility indices use a single cardinal scale to describe the disutility of ill-health as perceived by the general population. Both EQ5D and QWB use the same scale - i.e., a 0-1 scale where 0 represents a state that is as bad as being dead and 1 represents the best of health, with the scale having interval properties (i.e., like the temperature scale from 0 -100 C). Such health utility indices normally obtain ratings of health states based on people's stated preferences. (Further details are given in Chapter 3.)

6.13 There are other types of scales, used to describe in a systematic and detailed way the severity of specific diseases - an example is the St George's Respiratory Questionnaire (see Chapter 3). Typically, disease-specific scales do not use the same end points as health utility indices, or have interval properties, or base their ratings on people's preferences - hence comparisons cannot be made between diseases.

Quantification of morbidity

6.14 The COMEAP quantification report gave estimates of the numbers of hospitalisations induced by air pollution. This is, however, a measure of health care process rather than a measure of morbidity. What we really want to assess is how much of a deterioration in health is caused by air pollution amongst those people for whom the deterioration is such that they need to be admitted to hospital. Unfortunately, however, there are no studies which assess directly how health-related quality of life changes with hospitalisation, so we have had to use indirect methods.

Quality of life before and during hospitalisation

6.15. Ideally, we would want to know how much better people's health-related quality of life would be in the absence of air pollution-induced hospitalisation. While the ideal evidence is not yet available, we can gather some indications from existing studies. In Chapter 4 (paragraph 4.16) we noted some results from a recent study (as yet unpublished: Ayres *et al*, personal communication) which measured the health of COPD patients in Birmingham. In this study, these COPD patients rated, on the EQ5D scale, their health-related quality of life before hospitalisation at about 0.5 on average (1 SD, from 0.2 to 0.8; these figures include those who survived the study period as well as those who died during it).[3] We expect that those at

---

[3] Another study, of COPD outpatients in Sheffield, also found that such people rated their health-related quality of life on the EQ5D scale at about 0.5 on average (Harper *et al*, 1997).

risk of hospitalisation due to the effects of air pollution may be in a similar state of health. It seems to us reasonable to suppose that these people would still suffer from their underlying respiratory diseases even in the absence of the air pollution. However, the acute morbidity which would trigger hospital admission should be relieved - so we would expect their health-related quality of life to rate higher than if air pollution triggered their admission to hospital.

6.16 There were no data available on quality of life during the deterioration which had resulted in admission. However, we suggest as rough bounds that, on the EQ5D scale, the deterioration could range from zero (i.e., no worsening of health, or a change imperceptibly small), to being as large as a deterioration from 0.8 to death (0). (The high estimate is based on the mean plus one standard deviation which will not encompass the full range of the data, so larger changes are possible. We expect, though, that for most of those at risk the true change in quality of life would probably fall into a much narrower range than this. Using the average "baseline" rating of 0.5 would narrow our range of possible deteriorations from no change to 0.5.)

6.17 There is no direct empirical data on what QWB ratings would be in the absence of air pollution-induced hospitalisation. There is, however, a study (Kaplan *et al*, 1984) suggesting that (non-hospitalised) COPD patients who underwent an exercise programme had a QWB score of about 0.6 ($\pm$ 1 SD 0.5-0.7). We take this (in the absence of better evidence) as our estimate of a baseline QWB rating of quality of life for people at risk of hospitalisation from air pollution.

6.18 Direct evidence on QWB scores during hospitalisation for respiratory disease is not available. In the absence of direct empirical evidence, we have estimated a figure of 0.47[4]. We use the difference between this figure of 0.47 and the score of 0.6 from Kaplan *et al* (1984), above, to produce an alternative measure of 0.13 (interval 0.03-0.23), on the QWB scale, as representing the deterioration in health due to air pollution-induced hospitalisation.

6.19 The above calculations can be used to derive figures for the numbers of quality-adjusted life years involved in air pollution-induced hospitalisation, by combining the reduction in quality of life with its duration[5]. The duration of hospitalisation for respiratory disease which might be attributed to air pollution is about 8 days (all ages) to 14 days (65+; see Table 5.1). A spell in hospital of this duration, therefore, represents about 2% to 4% of a life-year. Combining this with the reductions in quality of life on the EQ5D scale (see paragraphs 6.15 and 6.16 above) suggests that hospitalisation involves the loss of between 0% and 3% of a quality-adjusted life year per spell. (We consider the EQ5D scale most appropriate for our calculation of quality-adjusted life years as it is the most commonly used scale in the UK. We discuss the use of the QWB scale further in the context of predicting willingness to pay (WTP) in paragraphs 6.41-6.42.) We consider that giving the estimate as a range from 0% to 3% best reflects the uncertainties. (A single "middle" estimate could be derived based on the average pre-hospitalisation EQ5D rating of 0.5, a mid-point between no change and a change of 0.5 - i.e., 0.25 - for the deterioration, and a mid-point of 11 days for duration - although the mid-points are not necessarily more likely than another point within the range. This would give a middle estimate of 0.75% of a quality-adjusted life year per spell.) Given the numbers of hospitalisations reported in the COMEAP quantification report, our range for total quality-adjusted life years lost due to hospitalisations for each pollutant would be:

- PM_{10} 0 - 315
- SO_2 0 - 105
- O_3 0 - 297

---

[4] We have used the standard QWB rating scales (see Annex 3A) to calculate a score. The numbers represent decrements compared with not having these problems.
    Mobility: In hospital (-.09)
    Physical activity: In bed, chair or couch (-.077)
    Social activity: No major role activity, needing help with self-care (-.106)
    Symptoms: Cough, wheezing or shortness of breath (-.257).
    Total score: 1 - .09 - .077 - .106 - .257 = 0.470.
It should be noted that the choice of the above descriptors is based merely on our judgement of those that would appear most appropriate from the standard QWB set, and does not represent the expressed views of patients themselves or their physicians, who might have chosen other descriptors with different values.

[5] Compare paragraphs 4.15-4.20.

It should be noted that the only loss of quality of life included in the above figures is that during a relatively short spell in hospital. We give the quality-adjusted life years gained per unit reduction in concentration of pollutant in Table 7.3.

6.20 There is much uncertainty relating to these calculations, as noted above. As with NHS costs (paragraph 5.8), the above assumes that the distribution of ages of people affected is the same as for respiratory patients in general. This may not be the case and duration of stay is dependent on age (average 14 days in the over 65s compared with 4 days in the under 65s). Moreover, there are also issues concerning age and other factors that are expected to affect people's preferences - as discussed in Chapter 4 and set out in Table 4.1. Unfortunately, we lack the data to assess the practical significance of such factors.

Other morbidity and quality of life effects

6.21 As noted at paragraph 6.5, there is likely to be a range of other types of morbidity caused by air pollution. Quality of life ratings have been attached to some of these (see: Johnson *et al*, 1996; Maddison, in press). Table 6.1 below sets out estimates of the QWB ratings for a day in various health states. Unfortunately, there is too little good epidemiological data on the frequency and duration of these effects, as caused by air pollution, to assess the aggregate numbers of quality-adjusted life years at stake - but as such data emerge ratings such as those in Table 6.1 may be useful. (Of course, better still would be direct assessments of quality of life and WTP by those at risk, and we have recommended further research on this.)

Valuation of morbidity

6.22 Just as the epidemiological measures of the precise impacts of pollution on morbidity are in some doubt, so too the tools to value these effects are not as well developed as in studies of the valuation of reducing mortality risks. Nevertheless, attempts have been made to put relative monetary values on the various states of ill-health and some of these estimates can, we believe, be read across into the analysis of the benefits of pollution control.

What should we value?

6.23 As noted in Chapter 5, there are a number of different aspects to the costs of air pollution-induced ill-health, as follows:

(i) private costs of dealing with illness - e.g., travelling to see a doctor, or buying eyedrops - and costs of trying to avoid ill-health through avertive behaviour (from buying cycling masks to moving house);

(ii) public (or third-party) costs of treating illnesses - e.g., time spent by GPs; spells in hospital; the cost of prescribed drugs;

(iii) loss of output - in the case of all but major ill-health this is generally a social rather than a private cost, given that people receive sick pay;

(iv) wider disutility costs of ill-health, both to the individual and to his/her family and friends - sometimes referred to as "pain, grief and suffering".

Chapter 5 deals with the first three of these, while here we deal with the fourth.

6.24 One difficulty is to disentangle these costs and avoid double-counting. For example, if people's WTP to avoid particular ill-health states is used as a measure of the fourth cost, but respondents have actually also reflected the first three as well in their WTP values, then adding estimates of the financial costs of the first three to the WTP values would involve double counting. However, as explained in paragraph 5.2, we believe that, with third parties paying the bulk of the financial costs, most people's own direct expenditures and financial losses for the first three will typically be small, at least in relation to their assessment of how the fourth cost affects them. Thus, estimates of WTP to avoid the disutility of ill-health are unlikely to be much "contaminated" by the other costs. (However, there may be groups, such as the self-employed, for whom this is not such a safe assumption.)

Stated preference

6.25 Much of the work undertaken to value the disutility of ill-health has been based on interviews or questionnaires. There are two different ways of presenting the different types of ill-health to people:

- where the ill-health is well known and understood, e.g., a cough, then direct questions can be asked;

- but, given people's ignorance of the full impact of many types of ill-health (e.g., of, say, COPD) respondents should generally have the symptoms associated with each illness e.g., degree of pain, immobility explained to them before being asked to consider how they respond.

6.26 There are, too, direct and indirect approaches to deriving monetary valuations of ill-health or the avoidance of ill-health:

- a direct approach is simply to pose different states of health-related quality of life and to ask people how much they would pay to avoid each state e.g., asking directly about WTP to reduce the number of days on which they suffer particular symptoms;

- an indirect approach is to obtain valuations in stages:

    - the first stage is to describe the morbidity using a quality of life classification system - i.e., using standard quality of life descriptors - and apply the quality of life utility ratings so that the morbidity can be calibrated on the 0-1 index;

    - the second is to assess WTP for movements between health states with different ratings - hence to attach values to increments or decrements of, say, 0.1 on the quality of life scale.

6.27 In such valuations, many of the factors listed in Table 4.1 will be as relevant to morbidity as in the case of mortality risks. So, for example, the responses people give may be affected by :

- whether a pollution context is specified;

- whether people feel that the risk is under their control and/or is someone else's responsibility;

- how long effects will persist; and

- whether, say, the disease is one regarded with particular dread (e.g., as with cancer).

Willingness to pay studies available

6.28 We have been assisted in our work by reference to various overviews of the morbidity valuation literature (Maddison, in press; Cropper and Freeman, 1991) and draw on them in our discussions below. It is clear that the literature addressing morbidity valuation is currently much more limited in scope than the risk-of-mortality valuation literature.

6.29 One approach, which has been used in the USA (see, e.g., Ostro and Chestnut, 1998), starts from estimates of the financial costs of ill-health and factors these up, given that they inevitably underestimate the full welfare cost. It has been suggested that multiplying the financial costs of ill-health by a factor of 2 would lead to a rough estimate of WTP - based on such evidence as there is of the relationship between them. In other words - using the partitions in paragraph 6.23 - it has been estimated that category (iv) will be double the sum of the other three categories.

6.30 However, this ratio will be different for different types of ill-health. We see no particular reason why the suffering caused by ill-health (on which WTP is based) should be systematically related to treatment costs - e.g., some serious illnesses may be untreatable and have low treatment costs. Also, since the financial costs of ill-health in the USA may well be higher than in the UK, such results may not be readily transferable. Therefore, we do not consider this an ideal approach. At best it might give some help in simply gauging whether estimates derived in other ways seem to be of the right orders of magnitude.

6.31 Such values have been reported in ExternE (European Commission, 1995). Their figures suggest that the WTP values for avoiding hospitalisation for respiratory or cardiovascular disease would be around £5,000. But these are based on US costs, and it is unclear whether these values would be applicable to the UK. Based on the NHS costs set out in Chapter 5, the value derived would be around £2,800 - about 40% less. However, we believe that any such estimates, based on factoring up financial costs, are highly uncertain.

6.32 Another approach has based WTP estimates on people's avertive behaviour. For example, Cropper and Freeman (1991) report that Chestnut et al (1988) studied the extent to which people bought in help - e.g., paying others to do work to their home or car - in order to avoid angina pain, which they might experience if they undertook such tasks themselves. Unfortunately, it is very difficult to assess how much of such expenditure should be attributed to the avoidance of angina, and how much to other benefits - e.g., better quality of the end result, or extra leisure time enjoyed - from paying someone else to undertake the tasks.

6.33 A direct approach has been to ask people to state their WTP to avoid extra days with specified symptoms. An early example is Loehman *et al*, (1979).

> Loehman *et al*, (1979) used a questionnaire which asked people to specify a maximum WTP - for example:
>
> "To avoid 1 day per year minor head congestion, the most I would pay is $0 / $0.50 / $1 / $2 / $10 / $15 / $50 / $120 / $250 / $1,000".
>
> Rather than questions being about named diseases, respondents were asked about symptoms which they might understand more clearly - e.g., they were asked about "severe shortness of breath" rather than, say, "asthma". Also included were severe cough, chest pain, and eye irritation. The study was undertaken in the Tampa, Florida area. The questionnaire was mailed to 1800 people selected randomly, with a return response of 404. As an example of the results, the median value put on avoiding a day of severe shortness of breath was $70 (at 1993 prices).

Subsequently, in the 1980s, a number of such studies (Rowe and Chestnut, 1985; Tolley *et al*, 1986; Dickie *et al*, 1987; Chestnut *et al*, 1988; reviewed by Cropper and Freeman, 1991; and Johnson *et al* 1996) were carried out for the US Environmental Protection Agency. They have used differing definitions of ill-health, and methods, so it is not straightforward to make direct comparisons of their results.

6.34 A more recent example of this type of study has been carried out in Norway (Navrud, 1997). One reason for interest in this study is that, being recent, it has been able to take advantage of earlier experience with WTP studies, which has led to the development of guidelines for best methodological practice (e.g., NOAA, 1993). Another is that the respondents are European - which is rare in such studies. A particular point of interest is that it shows how familiarity with a disease can influence responses - asthmatics valued the avoidance of an asthma attack-day twice as much as non-asthmatics. Some further notes on the methods and results are in the box below.

> In Navrud's study a representative sample of 1,009 Norwegians were interviewed in person by the Norwegian Gallup Institute (with 91.2% of the sample answering the questions) in April 1996. They were asked about WTP to avoid days of coughing, sinus congestion, throat congestion, eye irritation, headache, shortness of breath, acute bronchitis, and asthma. Half were asked about avoiding one more day in the next 12 months, over and above their normal experience (based on the previous 12 months), while the other half were asked about avoiding 14 more days. Air pollution was not explicitly mentioned to the respondents. As well as being asked their maximum WTP, they were reminded that they needed to think about the impact of payment on their household budget (to try to avoid exaggerated bids) and the ways in which the symptoms would affect them (to try to avoid understated bids). They were allowed to change their answers (although only 11% did). They were also asked about WTP to avoid days with a combination of three symptoms. Some results are reported here - these are the values (1996, converted to £s[6]) of avoiding one extra day:
>
> - throat congestion — £ 8
> - coughing — £ 8
> - eye itching — £10
> - headache — £14
> - sinus congestion — £15
> - acute bronchitis — £16
> - shortness of breath — £21
> - throat & sinus congestion & coughing — £29
> - asthma attacks (non-asthmatic respondents) — £45
> - asthma attacks (asthmatic respondents) — £93

6.35 A recent study in Hamilton-Wentworth, in southern Ontario (Diener et al, 1997) has explored the values attached to a range of outcomes of air pollution. The outcomes in the study were health effects (deaths and hospital admissions), poor visibility, black fallout, and bad odour. The study used conjoint analysis to assess WTP for reductions in these effects (through seeking preference rankings across alternatives in which these factors were varied, and where property tax or rent was also varied). However, rather than assess individual WTP for a reduction in personal exposure to risk, the questions were framed in terms of reductions across the region (and in terms of certain reductions, rather than risk reductions) so that individuals had to make their own assessment of how much less they or their household would suffer. This study is of particular interest as it directly asks about WTP to reduce deaths and hospitalisations due to air pollution. The average household's WTP was (Canadian) $40 per month to reduce deaths and hospitalisations in the region by 1 and 6 per month, respectively, (and the average household was further willing to pay more to achieve reductions in the other nuisance effects). Unfortunately for our purposes, the health effects were grouped together and so no separate valuations of deaths and hospitalisations were obtained. Moreover no overall values are given, aggregated up to the level of a total community WTP for reductions in the health effects. Partly this may have been because the results were from households who were not entirely typical of the community in general - e.g., respondents were more likely to have university degrees and be homeowners, and may have been more concerned about air quality and its effects than the average household. (In turn this reflects the way in which the survey was undertaken - it was mailed to 1900 households, and the response rate was about 30%.) Although this study does not give us the WTP figures which we would ideally like, it does seem to offer a potentially useful methodology which could be replicated.

6.36 Values attached to avoiding chronic bronchitis have been assessed directly in the USA. Viscusi et al (1991) estimated WTP to avoid a 1/100,000 risk of chronic bronchitis to be about $9 - implying aggregate WTP of about $0.9m to avoid one case (see box below). Chronic bronchitis is primarily a disease of smokers. This study used a sample of people in a shopping mall with varying smoking histories, including non-smokers who are not at significant risk of developing chronic bronchitis. However, there is no *a priori* reason to suppose that non-smokers

---

[6] Converted at £/NOK exchange rate at 20/8/98.

responding to questions which specified a hypothetical risk would give substantially different answers to smokers. In an update to that study involving people who, having a relative with chronic bronchitis, might be expected to know much more about chronic bronchitis and its effects, Krupnick and Cropper (1992) found that familiarity with the condition increased WTP.

> The study reported by Viscusi et al (1991) was undertaken in a shopping mall in Greensboro, North Carolina. Respondents (389) sat at a computer and responded to questions on the screen. They provided background information on their lifestyle, and thus, on factors which would affect their risks of suffering chronic bronchitis and also death in a road accident. Various health implications of chronic bronchitis were described (although most of the respondents would have had little prior or direct knowledge of the condition). They were asked to choose, if they had to move home, between different (fictitious) locations with differing (but lower than current) risks of chronic bronchitis and death in a road accident. They were also told how the locations differed in the cost of living. From the expressed preferences between the locations, the researchers inferred people's willingness to trade increased cost of living for reduced risk of chronic bronchitis, and other trade-offs such as risk of chronic bronchitis for road fatality risk. The mean WTP to avoid chronic bronchitis (aggregated up to the equivalent of avoiding one case) was $883,000. And the mean willingness to trade-off the two risks implied that avoiding chronic bronchitis was worth 68% of the VoSL from avoiding a road fatality.

6.37 Most of the available studies have been undertaken in the USA. Their coverage of types of ill-health is limited to certain symptoms. Ideally we need a way of overcoming the limitations imposed both by the paucity of data and the fallibility of some of the studies, while at the same time providing a basis for estimating the benefits of avoiding other illnesses, not covered in the original surveys.

6.38 The most fruitful approach to improving on the existing literature seems to us to use what we have described in paragraph 6.26 as the "indirect" approach. We believe it would be helpful to try to generalise the data relating to specific health conditions by translating each health condition into a quality of life measure, as suggested by Johnson et al (1996) and developed by Maddison (in press) in the air pollution context. The quality of life measure used in these studies was the quality of well being (QWB) scale plus details of the number of days on which the morbidity effect occurred. Once this is done statistical tests can be applied to see whether a general relationship between WTP, QWB scores and days can be established. If a relationship can be obtained then, within the plausible boundaries of the analysis, it should be possible to interpolate values for health conditions which have not been directly assessed, but which can be given QWB scores. Some results derived using this approach are set out in Table 6.1.

The method used to derive the results is described in the box below.

Table 6.1   Indirectly derived WTP values to avoid various health conditions

HEALTH CONDITION	QUALITY OF LIFE (QWB RATING)	WTP TO AVOID 1 EXTRA DAY (1996 £s)	WTP TO AVOID 14 EXTRA DAYS* (1996 £s)
Phlegm	0.830	15	56
Eye irritation	0.770	22	81
Cough	0.743	26	96
Respiratory symptoms	0.743	26	96
Minor restricted activity day	0.743	26	96
Chest discomfort	0.701	35	126
Major asthmatic attack	0.622	62	225
Acute bronchitis	0.622	62	225
Bed disability day	0.605	70	256

* The 14 days need not be consecutive, but could be any extra 14 days of symptoms in the next year.

> **Methodology used in deriving results in Table 6.1**
> Maddison undertook a meta-analysis using results obtained from previous studies. Data were taken from six earlier studies in which WTP to avoid extra time in impaired health had been derived - the studies mentioned in paragraphs 6.33 and 6.34. Information was available on the nature of the impaired health state and WTP to avoid a (specified) number of extra days in the state. The states were assessed in terms of health-related quality of life and given estimated ratings (using the QWB scale). There were 68 different states with WTP values attached. These were treated as observations and the related QWB ratings and numbers of days were treated as explanatory variables in a regression analysis. The best fit equation - from which the above results were obtained - was:
>
> $\ln(WTP) = 1.76 - 4.80 \ln(QWB) + 0.49 \ln(days)$.
>
> This equation gives results in £ at 1993 prices. The results have been revalued to 1996 prices using the retail prices index (+8.5%).
> ($R^2 = 0.76$. The t statistics are 5.25 for the constant term; 6.27 for the coefficient on $\ln(QWB)$ and 11.47 for the coefficient on $\ln(days)$.)

6.39 The health conditions in Table 6.1 assessed by Maddison include many of those with which we are concerned. It is interesting to note that the WTP to avoid given losses of quality of life seems to be subject to increasing valuation as the health states worsen - so that a decline from 0.7 to 0.6 is valued more highly than a decline from 0.8 to 0.7. (Such a finding of increasing WTP ought, in principle, to have implications for the way in which quality-adjusted life years are aggregated - i.e., the starting/finishing point of any change ought not to be ignored, as an increment from a worse health state would have a greater value than the same increment from a better health state.)

6.40 A finding that emerges strongly from these data is that WTP per day decreases with extra days - so that avoiding an extra day of ill-health on top of ill-health that has occurred on 30 days has a lower value than avoiding an extra day of ill-health on top of ill-health that has occurred on only 7 days. This finding seems to contradict that reported in relation to the rating of EQ5D health states. The Measurement and Valuation of Health Group (1995) actually report separate quality of life ratings for states lasting one month, one year, and ten years, with given decrements in health having higher values the longer the duration. An explanation of this apparent discrepancy might be that while the marginal disutility of a day's ill-health increases with duration, at the same time the marginal utility of money also increases, so that people are unwilling - or unable - to pay more and more to avoid extra days in the health state in question as, eventually, ability to pay becomes constrained.

Relevant valuation estimates

Hospitalisation

6.41 Our particular need is to derive a value for respiratory hospital admissions. There are no studies directly assessing how admission itself affects quality of life. But we suggest that, as an interim measure, a reasonable approach is to use an appropriate scoring system for health-related quality of life, and to read off an estimate of WTP from the analysis described above at paragraph 6.38 and Table 6.1. It should, however, be borne in mind that the studies used to derive the statistical relationship between WTP and QWB scores mainly concerned more minor morbidity effects than hospital admissions. Our range of changes in quality of life goes beyond the range covered by the data used in the modelling underlying Table 6.1 and it is unclear whether the relationship holds outside the original range of QWB scores. (The lowest rating in the modelled data was only 0.58, for severe angina, whereas we wish to value a lower rating, of 0.47, and so extra uncertainty arises from this.)[7]

6.42 Our assessment of the effect of hospitalisation on QWB scores is set out above at paragraphs 6.17 to 6.18. We also note at paragraph 6.39 that a loss of, say, 0.2 on a QWB scale seems worse from a low state than from a high state, and from paragraph 6.40 that the value does not increase linearly with the number of days. Based on these considerations, we suggest a low estimate of reduction in QWB score of 0.03 (from 0.5 to 0.47), and a high estimate of 0.23 (from 0.7 to 0.47). We also take a low estimate of duration of hospitalisation of 8 days and a high estimate of 14 days. From the model underlying Table 6.1, we then get the following results:

QWB SCORES

	0.7	0.5	0.47	DIFFERENCE
8 days	-	£488	£656	£168
14 days	£127	-	£862	£735

6.43 Thus, our low estimate of WTP to avoid deterioration in health due to air pollution-induced hospitalisation is £168, and our high estimate is £735 (at 1996 prices). We consider that giving the estimate as a range from £170 to £735 best reflects the uncertainties. A single "mid" estimate could be derived using the mean QWB score of 0.6 and a mid-point of 11 days for duration (although this is not necessarily more likely than another duration between 8 and 14 days). This would give an estimate of WTP of about £530 (1996 prices).

6.44 There are many uncertainties involved with these estimates. One concerns whether the QWB quality of life ratings, which we used to derived the estimates of WTP, are compatible with the EQ5D ratings, with which we estimated quality-adjusted life-years. The QWB and EQ5D both aim to measure the same thing (i.e., the disutility of loss of health), both are interval scales, and both scales use the same definitions of the 0 and 1 end points. Therefore, in theory the same change should take the same utility value on each scale. However, in practice there are differences. For example, the EQ5D scale acknowledges that some disease states could be regarded as "worse than death" and allows negative ratings (and in surveys people do give negative ratings - see Annex 3A). This is not an option on the QWB scale. In addition, the QWB scale does not appear to give low values of 0.2 or 0.1 (see footnote to paragraph 6.41). Some differences may be due to differences in the specified duration of the quality of life states when obtaining the ratings from surveys of the general population, but some will be due to the design of the scale. As mentioned previously, the EQ5D is most commonly used in the UK but an equation relating WTP to EQ5D scores was not available.

---

[7] Moreover there are some unusual aspects to the QWB scale. The minimum scores for mobility, physical activity, social activity and symptom complexes give, in combination, an overall score of 0.36 (severe burns) or 0.32 (loss of consciousness) or, for death, a score of 0. It seems odd that a QWB score of 0.2 or 0.1 is impossible. Also, "cough, wheezing and shortness of breath" is ranked as less severe than a symptom complex including "discomfort in the chest" but can in fact be severely disabling.

6.45 The difference that air pollution makes to quality of life is also uncertain. Assessing the quality of life of being hospitalised is only part of the picture. For example, we lack direct information on the state of health of those hospitalised before and after hospitalisation, and the size of the change in health status that can be attributed to air pollution. The size of the reduction in quality of life has a significant impact on our estimates, so it is important to try to be accurate. Assessments at different points in time from prospective studies are needed to help fill these gaps in the empirical evidence.

6.46 Another uncertainty concerns whether valuation is proportional to duration and severity of ill-health. If there is increasing marginal disutility of ill-health then extra days in ill-health, or in worse health, for someone already suffering from poor health, or for a long time, will be worse - i.e., carry a higher value - than for someone who has suffered less, or for a shorter time. Conversely, decreasing marginal disutility of ill-health will have the opposite effect. Again we lack data on the cumulative nature of the ill-health suffered, the duration of hospitalisations attributable to air pollution, and on how long prior and subsequent air pollution-induced morbidity lasts.

Other morbidity effects

6.47 The COMEAP report was not able to quantify any more minor effects of air pollution, although some of the evidence was described. Two morbidity outcomes which we have considered for use in sensitivity analysis are bronchodilator use and cardiovascular admissions. We also understand that data on associations between GP consultations and levels of pollution may be published in the near future.

- Bronchodilator use may be helpful as an indicator of more minor effects in patients with asthma or COPD. We suggest that WTP to avoid the need for such use would be less than WTP to avoid hospitalisation, but might be similar to WTP for avoiding coughs and respiratory symptoms, or even acute bronchitis and asthmatic attacks this suggests a value of between £26 and £62 to avoid 1 extra day per year.

- Morbidity which prompts a visit to the GP may be more serious than that which only increases bronchodilator use (but, again, less serious than that involved in hospitalisation). We suggest that the WTP to avoid the need for such a visit might be similar to the WTP for avoiding major asthmatic attacks or bed disability days. This suggests a value of between £62 and £70 to avoid 1 extra day per year.

- It is extremely difficult to value cardiovascular hospital admissions since there can be a very wide variation in the degree of morbidity - from transient chest pains to major long-term incapacity. There is relatively little understanding of the type of circulatory disease affected by air pollution. However, our estimate of the value of avoiding being hospitalised does not depend on the type of morbidity suffered. Rather it is based on the degree to which activities of daily living are limited, and we suspect that this will be similar. (If length of stay is a little longer, then our method would give a slightly higher value.) We lack data on other dimensions of health. Thus, our best estimate would be similar to that for hospitalisation with respiratory disease.

- We note that there is a little information on WTP to avoid angina symptoms (see paragraph 6.32 above). However, we think that the WTP information is not robust and there is essentially no good epidemiological data on the effects of air pollution on cardiovascular symptoms.

6.48 There is little evidence around the valuation of chronic morbidity. However, one useful study is that by Viscusi *et al* (1991) which estimated the value attached to avoiding chronic bronchitis in the USA. WTP to avoid a 1/100,000 risk of chronic bronchitis was about $9 - implying aggregate WTP of about $0.9m to avoid one case. This might form the basis for valuing some of the chronic effects of air pollution, as and when such evidence emerges.

Conclusions

6.49 Comparing the above estimates with the costs to the NHS derived in Chapter 5, we can see that the WTP to avoid the disutility involved would be much less than double those figures. Thus, the "cost of illness times two" approach discussed earlier (paragraphs 6.29 - 6.31) receives no support.

6.50 Unfortunately, we lack reliable evidence on morbidity other than the number of hospitalisations. However, given the greater numbers of people affected, the sums may be as large as (or larger than) those we have estimated for hospitalisations.

6.51 Comparison with the mortality effects suggests that the burden of mortality far outweighs that of morbidity (given that the numbers of deaths and hospitalisations are similar).

6.52 Nearly all our data refer to the acute effects of air pollution, and there is a serious lack of information about the chronic effects. We feel that this is one of the biggest and most urgent gaps in the evidence base.

6.53 We recommend more direct assessments of WTP. We think it would be useful if studies of the effects of air pollution were to use standard descriptors of outcomes in quantifying the impacts. Preferably these standard descriptors should be based on (or at least be clearly translatable into) those used for health utility indices. More direct WTP studies should also be undertaken based on the valuation of outcomes characterised using the same health utility index descriptors.

6.54 Finally, we have been made aware of an EC project which has been undertaken over the past couple of years (Dubourg, 1998). This is assessing directly, in various EC countries including the UK, WTP to avoid certain air pollution-related health outcomes, such as hospital admission, emergency-room visits, being confined to bed, a day of coughing, and a day of eye irritation. When the results of this work are available, they may go some way towards meeting the needs for further evidence which we have highlighted.

References

Chestnut, L.G., Colome, S.D., Keller, L.R., Lambert, W.E., Ostro, B. and Rowe, R.D. (1988) *Heart disease patients' averting behavior, costs of illness, and willingness to pay to avoid angina episodes.* Washington DC: US Environmental Protection Agency.

Cropper, M.L. and Freeman, A.M. (1991) Environmental Health Effects. In: *Measuring the Demand for Environmental Quality.* (Braden, J.B. and Kolstad, C.D. eds) Amsterdam: North Holland.

Department of Health (1998). *Health Survey for England 1996.* London: The Stationery Office.

Dickie, M., Gerking, S., Brookshire, D., Coursey, D., Schulze, W., Coulson, A. and Taskin, D. (1987) Reconciling averting behavior and contingent valuation benefit estimates of reducing symptoms of ozone exposure. In: *Improving Accuracy and Reducing Costs of Environmental Benefit Assessments.* Washington DC: US Environmental Protection Agency.

Diener, A.A., Muller, R.A. and Robb, A.L. (1997) *Willingness-to-Pay for Improved Air Quality in Hamilton-Wentworth: A Choice Experiment.* Working Paper No 97-08. Hamilton, Ontario, Canada.: Department of Economics, McMaster University.

Dockery, D.W. and Pope, C.A. (1994) Acute respiratory effects of particulate air pollution. *Annu. Rev. Public Health Med.* **15**; 107-132.

Dubourg, W.R. (1998) *Benefits transfer and the economic evaluation of environmental damage in the European Union: with special reference to health. An Overview.* London: University College, CSERGE.

European Commission (1995). *ExternE - Externalities of Energy. Volume 2. Methodology.* Luxembourg: Office for Official Publications of the European Communities.

Harper, R., Brazier, J.E., Waterhouse, J.C., Walters, S.J., Jones, N.M.B. and Howard, P. (1997) Comparison of outcome measures for patients with chronic obstructive pulmonary disease (COPD) in an outpatient setting. *Thorax* **52**: 879-887.

Johnson, F.R., Fries, E.E. and Banzhaf, H.S. (1996) Valuing morbidity: an integration of the willingness to pay and health status index literatures. (Working Paper) Durham, North Carolina: Triangle Economic Research.

Kaplan, R.M., Atkins, C.J. and Timms, R. (1984) Validity of a Quality of Well Being scale as an outcome measure in chronic obstructive pulmonary disease. *J. Chronic Dis.* **37**: 85-95

Kruppnick, A.J. and Cropper, M.L. (1992) The effect of information on health risk valuations. *J. Risk Uncertainty* **5**: 29-48.

Loehman, E.T., Berg, S.V., Arroyo, A.A., Hedinger, R.A., Schwartz, J.M., Shaw, M.E., Fahien, R.W., De, V.H., Fishe, R.P., Rio, D.E., Rossley, W.F. and Green, A.E.S. (1979) Distributional analysis of regional costs and benefits of air quality control. *J. Environ. Econom. Manage.* **6**: 222-243.

Maddison, D.M. *Valuing health effects of air pollution.* Edward Elgar. (In press)

Measurement and Valuation of Health Group (1995). *The Measurement and Valuation of Health*, York: University of York, Centre for Health Economics.

Navrud, S. (1997) Valuing health impacts from air pollution in Europe: new empirical evidence on morbidity (Working Paper). Norway: Agricultural Unviversity of Norway.

NOAA Panel (1993) Report of the NOAA Panel on contingent valuation. Washington DC. Federal Register 58(10), 15 January 1993.

Ostro, B. and Chestnut, L. (1998) Assessing the health benefits of reducing particulate matter in the United States, *Environ. Res.* **76**: 94-106.

Rowe, R.D. and Chestnut, L.G. (1985) *Oxidants and asthmatics in Los Angeles: a benefits analysis. Report No EPA-230-07-85-010.* Washington DC: US Environmental Protection Agency Office of Policy Analysis.

Tolley, G.S., Babcock, L., Berger, M., Bilotti, A., Blomquist, G., Fabian, R., Fishelson, G., Kahn, C., Kelly, A., Kenkel, D., Krumm, R., Miller, T., Ohsfeldt, R., Rosen, S., Webb, W., Wilson, W. and Zelder, M. (1986) Valuation of reductions in human health symptoms and risks. Report for USEPA Grant No CR-811053-01-0. Washington DC: US Environmental Protection Agency.

Viscusi, W.K., Magat, W.A. and Huber, J. (1991) Pricing Environmental Health Risks: Survey Assessments of Risk-Risk and Risk-Dollar Trade-Offs for Chronic Bronchitis. *J. Environ. Econom. Manage.* **21**: 32-51.

Chapter 7
Summary and Conclusions

7.1 We stated in the introduction that this report would advise on how best to reflect the importance of the health effects of air pollution when balancing the advantages and disadvantages of restricting activities which generate pollution. The subsequent chapters explored several aspects of this, guided by our terms of reference. This chapter summarises the progress the group has made and discusses what more needs to be done.

7.2 This report is very much a first step in a difficult area beset by extensive uncertainty. The amount of empirical valuation work specific to the air pollution context is very limited and there was insufficient time for work to be commissioned and completed during preparation of the report. We have provided guidance where we can to assist with current policy development but hope that one of the benefits of our report will be the identification of areas in which further work should substantially improve the approach to the assessment of the benefits.

7.3 This report is mainly concerned with methodological issues. Our aim has been to provide building blocks to be used in a subsequent analysis of the UK National Air Quality Strategy (NAQS) which will be published separately. It is in this subsequent analysis rather than the current report that the actual benefits of any proposed reductions of air pollutants will be discussed. This analysis will not address the "total health costs of pollution", but rather incremental changes in pollution, to which our results can be applied. Our report advises on how to assess the benefits of reductions in air pollution, and how to value them from the perspective of the potential beneficiaries. The costs of bringing about reductions in air pollution will, we know, be considered separately, in the context of the NAQS. The costs of any reductions may fall on a different group of people than those who would benefit. Indeed it is a general principle of government policy that the polluter should pay - not the beneficiary. We do not consider that this poses any difficulties for our assessment.

7.4 Key issues arising from the previous chapters are highlighted and discussed below.

Quantification of health effects

7.5 The starting point for the assessment of the benefits of reductions in air pollution is the quantification of the health effects. This issue was addressed in the recent COMEAP report on the quantification of the health effects of air pollution. Strict criteria were used to determine where the evidence was robust enough to be used for quantification. It was concluded that the effect of PM_{10}, sulphur dioxide and ozone on all-cause deaths brought forward and on respiratory hospital admissions could be quantified.

7.6 The nature of these health effects is crucial to much of the further discussion in this report. The deaths are thought to occur mainly in the elderly who already have advanced lung or heart disease and to be brought forward by weeks or months but probably not years. We do not believe that the risks of such deaths are regarded by the population as a whole (including those affected) in the same way as, say, the risks to young adults in full health of dying in a road accident. Therefore, we have paid special attention to the particular circumstances of the health effects of air pollution.

7.7 Unfortunately, there is limited understanding of the exact characteristics of the people affected by air pollution as the studies on which we depend have mostly used routinely collected population statistics rather than details about particular individuals. In particular, the size of loss of life-expectancy is unknown although, as stated above, it is thought to be quite small. The COMEAP report recommended further research on this very important issue as well as on groups at special risk such as the elderly and chronically sick. We support these recommendations as further knowledge about the age, state of health and life expectancy of those affected is important for the purposes of economic appraisal.

7.8 Quantifying acute mortality and respiratory hospital admissions alone for just three pollutants is likely to underestimate the full health effects of air pollution. There is evidence for effects of these pollutants on other health outcomes and for effects of nitrogen dioxide and carbon monoxide although this is weaker and there is more uncertainty about the quantitative impact. However, it is still useful to have a rough idea of the potential contribution of these effects provided the uncertainties are always clearly stated. For this reason, we have discussed some options for sensitivity analyses in Chapter 2, based on the COMEAP report and discussions with some individual members of COMEAP. We also emphasise that the health effects of benzene, 1,3-butadiene and lead have not been covered in this report.

7.9 The effect of air pollution on cardiovascular disease is less well understood than that on respiratory disease. The estimated deaths brought forward include cardiovascular deaths but it is unclear what types of heart disease are involved or, for example, whether patients with underlying disease but no symptoms before their first heart attack are affected by air pollution. Cardiovascular admissions have not been quantified although a sensitivity analysis on the inclusion of cardiovascular admissions is suggested. In contrast to respiratory disease, there is no range of studies of the effects of the pollutants on cardiovascular symptoms in particular patient groups to support the interpretation of the studies of more severe endpoints (except for chamber studies of the effect of carbon monoxide on angina). We suggest that the further research mentioned in paragraph 7.7 above should include the effects of air pollution on heart disease.

7.10 It is known that air pollution has effects on more minor outcomes than hospital admissions and deaths. Potentially, this could affect a much larger number of people. This is the sort of issue which could be explored in sensitivity analyses. However, this is difficult because the size of the relevant disease group in the population and the baseline rate of the relevant symptoms in that group need to be defined and this may not be possible. The COMEAP report recommended further studies of the effects of air pollutants on outcomes other than death and hospital admissions. We support this and suggest that it would be helpful if such studies could consider the form in which baseline data are collected when defining their study subjects and symptom reporting. This would be of considerable assistance when assessing the overall public health impact and the economic aspects of such an impact.

7.11 One of the largest uncertainties in quantifying the overall health effects of air pollution is the possible extent of any chronic effects. The COMEAP report considered the evidence was not suitable for quantification purposes but noted that the chronic effects could be substantial. Determining the public health impact of the chronic effects is not straightforward - several important pieces of information (e.g., effect on different ages, lag time to effect) are needed to compare the effect on a population over time with and without a particular change in pollution. These are currently lacking and appropriate assumptions are not always obvious. Some work is underway in the UK estimating possible effects on life expectancy under various assumptions but this is still preliminary. In the time available to the group, we were not able to fully explore all these issues. The COMEAP report recommended future work on the impact on health of long-term exposure to current levels of air pollutants. We agree this is important. It would be useful if such work could include investigation of effects on people of different ages and of the lag time from exposure to effect. Continued work on methods of estimating the public health impact would also be helpful.

Assessing the importance of the health benefits - our approach

7.12 Once the health effects have been quantified, they need to be compared with the other effects of a proposed policy. In Chapter 3, we discussed a range of possible approaches to doing this. This group is advising specifically on the health benefits but we took the view that some consideration of the advantages and disadvantages of the overall approaches to policy appraisal was necessary so that we could express the health benefits in an appropriate form.

7.13 The approaches vary as to how much emphasis is given to the views of individuals, experts or society. The views from these different perspectives may all be quite different. For example, experts might consider that risks of the same magnitude should be treated the same whereas individual members of the public might consider that, if one of the risks was involuntary (as in the case of air pollution) then it was of more concern. There is no simple rule as to which perspective *should* be used. However, we have tried to ensure that the different perspectives are clearly stated so they can be taken into account as policymakers see fit.

7.14 Whatever approach is used, it is still important to understand the public's perception of air pollution risks as this influences the implementation of pollution control policies and provision of health advice. More research is needed on (i) the levels of knowledge and understanding people currently hold on the causes and effects of air pollution, (ii) what extra knowledge people might need to make informed decisions about their personal health protection and (iii) what processes of deliberation and information provision might help people to construct stable and consistent preferences regarding air pollution risk reduction.

Multi-criteria analysis

7.15 This approach expresses the various effects (particular health outcomes, healthcare costs, other costs etc.) of different policy options in terms of weighted ranks or scores as explained in paragraph 3.47. Most policy decisions will involve consideration of multiple factors in an informal way. Some forms of multi-criteria analysis were designed to help structure these decisions. It can also be used to structure the views of the public. Multi-criteria analysis can include information derived from cost-effectiveness analysis and cost-benefit analysis so it is not necessarily an alternative to these approaches. Although multi-criteria analysis will not show whether the costs exceed the benefits or vice-versa (unless cost and monetary valuation data are included), it can be useful in clarifying people's views about the trade-offs between different outcomes and to check whether the priorities suggested by willingness to pay (WTP) studies are in line with these. We suggest, therefore, that consideration is given to some future multi-criteria analysis work as a complement to the other approaches.

Cost-effectiveness

7.16 Cost-effectiveness analysis compares options against a fixed budget or against a fixed outcome. It does not show whether the options are worthwhile in terms of costs and benefits. A simple version is to compare costs per life saved of different programmes. However, as mentioned above in paragraph 7.6, this does not take account of the fact that people have different attitudes to deaths in different contexts or that the programmes may have other benefits. A form of cost-effectiveness widely used in the health service is to compare costs per quality-adjusted life year gained. In its current form it treats the loss of 1 quality-adjusted life year the same for people of all ages and for all diseases and contexts. Willingness to pay studies show that individuals do regard similar effects differently for different ages, that some diseases worry people more than others and that the context of the risk has an influence. Quality-adjusted life years could be adapted to take these into account. Whether they ought to be so adapted raises ethical issues which perhaps ought to be debated. For example, while people may place a higher value on avoiding a specific loss of health when the cause is cancer than when it is something else, when making personal decisions, they may feel that the NHS should not differentiate in this way.

7.17 There are various quality of life measures available. In general terms, the most appropriate scale to use in the UK would be the EQ5D. This has only just started to be used to look at chronic respiratory disease. There are no studies of the effect of air pollution on quality of life using this system (or any other). The EQ5D is a utility index covering the whole range from full health to death and is widely applicable. However, it may be insufficiently sensitive to detect small changes. A disease-specific questionnaire may be much more sensitive for specific aspects of health but may not cover the full range and is not applicable to other diseases. Unfortunately, very little work has been done to relate these types of measures. We believe more work should be done on the quality of life of patients with respiratory and cardiovascular disease and effects of air pollution on this using both general and disease-specific measures.

Cost-benefit analysis and monetary valuation

7.18 For a full cost/benefit analysis, costs and benefits need to be compared on the same metric. This means in practice that the benefits need to be expressed in monetary terms. This is done by finding out people's WTP for incremental reductions in the relevant risks.

7.19 We are aware that monetary valuation is viewed by some people with unease. This is particularly the case when the term "value of a statistical life" or "value of prevention of a statistical fatality" (VPF) is used as this can be perceived as saying "a life is only worth so much". We emphasise that this technique is not valuing individuals' lives as such but rather valuing small reductions in a risk of death. People are also concerned with the validity of the techniques used to determine WTP. Inferring WTP from people's behaviour (revealed preference) relies on finding a situation where people's behaviour is clearly related to the relevant risk. Contingent valuation asks people more directly about their WTP. People may be unfamiliar with this type of question and may be confused by the way questions are asked or what the risks mean. We have discussed these issues and concluded that well-conducted studies do give an adequate indication of the broad order of magnitude of the values people place on safety improvements.

Our approach

7.20 Given the complexity of the issues involved, the most fruitful approach in the longer term may be to use the different approaches alongside each other and to cross-check the results obtained. For this report we examined monetary valuation further in the air pollution context as the most appropriate technique to use for the purposes of a full cost-benefit analysis. The group believes that techniques developed by economists in several countries can be used to provide broad estimates of the monetary values which might be attached to improvements in health and reductions in the risk of dying. In addition, if air pollution control policies are to be pursued in part to achieve improvements in public health (in fact health benefits are a major proportion of total benefits), then it is inevitable that comparisons with other health improvement policies will be made. We therefore also gave further consideration to quality of life and life expectancy as this will allow comparisons with other health interventions. It is quite possible that there are other more cost-effective ways of improving the health of those with respiratory disease than by reducing air pollution (such as reductions in cigarette smoking) and policy-makers should be aware of this. However, it may not be straightforward to transfer resources allocated to improving the environment to healthcare and the questions being addressed here and in healthcare are different. Given that reduction of air pollution is being considered as a way of improving the environment, the key question here is whether the costs of doing this are justified by the benefits which would be gained. This report should contribute to answering that question.

7.21 We note that, on the one hand, environmental policy-making tends to use cost-benefit analysis which aims to maximise aggregate net benefits based on the population's preferences. On the other hand, health-care policy-making tends to use cost-effectiveness analysis which aims to maximise health gains in the population within budgetary constraints. We considered combining the WTP and quality-adjusted life year approaches as this would integrate the environmental and health-care policy approaches and reduce apparent inconsistencies. Simplistically, this could be done by dividing the VPF for a particular risk by the number of quality-adjusted life years lost. This has significant drawbacks - the two approaches, at least in their current form, are founded on different assumptions (see 7.16) and people's WTP to reduce their risk of death is not solely related to the numbers of years they might lose. There has been very little direct investigation of people's WTP for gains in quality of life and life expectancy. This issue is not specific to the air pollution context, but we consider it deserves further thought and wider discussion.

7.22 Final decisions in the areas of pollution control will always fall to a process of political decision-making based on a range of evidence. We believe it is better if decisions are informed by evidence of the likely balance of such costs and benefits as can be readily quantified. It is important to try and express the health benefits in monetary terms if possible because the health benefits form a major part of the overall benefits of reductions in air pollution. However, monetary valuation can be complemented by evidence from the other approaches we have discussed. There may of course be debate about the use of some of the approaches (paragraph 7.19) but we emphasise that they are only part of the story and do not provide the only evidence on which a decision will be based.

Benefits of reduced mortality risks

7.23 We paid particular attention to the consideration of the benefits of reduced mortality risks as these are likely to dominate the overall benefits.

7.24 We first considered whether we could estimate the numbers of quality-adjusted life years lost due to deaths brought forward by air pollution. This is uncertain - as noted above, the loss in life-expectancy is not known for certain although it is likely to be from a month or less to a year. In addition, there are relatively few studies of quality of life in patients with advanced respiratory disease. Preliminary data suggests a quality of life of about 0.4 (± 1 Standard Deviation (SD) 0.2 to 0.7)[1] in COPD patients with advanced disease (the group at risk). Thus, about 0.02 to 0.7 quality-adjusted life years might be lost per death brought forward. Even less is known about air pollution and cardiovascular disease.

7.25 We then considered whether there were any studies of WTP to reduce the risk of death being brought forward by air pollution. There are, of course, WTP studies available in other contexts. However, we considered that WTP studies should relate as precisely as possible to the population at risk. The relevant population might in some circumstances be the population as a whole; in others just those over 65; in others possibly just smokers. In this case, we needed to know the WTP of the groups most at risk from deaths associated with air pollution. We did not have these studies. This is obviously not an ideal situation but there was no time for work to be commissioned in the timescale of preparation of this report.

7.26 We could have left the matter there. However, we were aware that, in the short term, guidance was needed for the purposes of the planned cost-benefit analysis of the air quality strategy. The group believed that it was possible, from general knowledge of the factors that affect people's WTP, to provide some indication of the likely size of the WTP estimates for the relevant groups of the population, were it possible to administer the necessary questionnaire. This is what we have done. Inevitably, this involves a substantial degree of uncertainty and we emphasise that this is very much a tentative interim suggestion. Empirical work on this question is vital.

"Baseline" Air Pollution VPF

7.27 The basic principle of our approach was to take a "baseline" VPF figure as a reference and then apply a series of adjustment factors to reflect the characteristics relevant to air pollution mortality risks. One option for the baseline VPF was to use the well-established WTP-based DETR VPF for road accident deaths adjusted upwards for air pollution risk context (e.g., involuntary risk). Other options were based on overviews of different WTP-based VPF figures in the literature. These various options all gave similar results in the region of £2m. We adopted this as the air pollution baseline VPF for people of average age and in average health. This then needs to be adjusted to take account of the fact that the people affected by air pollution are not of average age or in average health.

---

[1] The use of the standard error rather than the standard deviation would be more appropriate but was unavailable. Our use of measures of variation is discussed in paragraph 7.54.

Adjustment factors

Age and impaired health state(reduced life expectancy and quality of life)

7.28 We have suggested adjustment factors for age and the impact of those affected already having an impaired health state for reasons other than air pollution. There is empirical evidence on the effect of age on WTP to reduce the risk of death. This shows a decline with age after middle age, perhaps in part because people are starting to accept the inevitable and may wish to make bequests to their children. Impaired health state can be regarded as a combination of reduced life expectancy and lower health-related quality of life. We expect reduced life expectancy to reduce WTP on theoretical grounds but it is unclear by how much. Theory and the relationship of WTP with age suggest WTP is unlikely to drop more than in proportion to life-expectancy. To adjust for both age and life expectancy may involve some degree of double-counting but the effect of age is not just about life expectancy and those in an impaired health state have an *extra* reduction in life expectancy below the average for their age. There is even less information on the expected impact of quality of life - apart from very severe health states being regarded as being as bad as death where WTP would be expected to be extremely low and a limited study showing WTP increased with increased quality of life (paragraph 4.78). Given the limited evidence, we have suggested the use of quality of life measures to scale down the expected WTP. We discuss this further in paragraph 7.35.

Baseline level of risk

7.29 We considered that the baseline level of risk from air pollution in the general elderly population was typically not so high as to require an adjustment to our baseline pollution VPF. However, this might need to be reconsidered once evidence is available on the level of risk in susceptible groups such as the elderly with advanced disease.

Acceptability of adjustments for wealth/income and other factors

7.30 Higher wealth or income is expected to lead to higher WTP. Therefore, in theory, an adjustment could be made if the group affected had an income significantly different from the average. However, this is one of the areas where it is generally thought appropriate that the principle that policies should not discriminate against the poor should override individual views which are affected by income. Uniform values of safety based on a representative sample of the population would usually be used. Therefore, we do not propose an adjustment factor for income. We recognise that similar arguments could be applied to some of the other adjustments factors e.g., if the young are cavalier about risks should this be acknowledged or overridden when developing policies? If people regard gains in life expectancy as less important when their quality of life is already poor, should this be overidden to ensure the most help goes to the weak? There is not such a clear consensus on - or much evidence about - these other factors. More generally, our approach is based on individual WTP and society as a whole may wish to allocate resources differently on grounds of altruism or equity. These wider concerns may be taken into account separately as part of overall policy development. We have set out our adjustment factors clearly so that they can be included or excluded as required.

Discount rate

7.31 Discount rates reflect the fact that a benefit some time in the future is less valuable to people than an immediate benefit. Discount rate adjustment of the baseline VPF does not, therefore, apply to acute mortality - where effects are immediate - but would apply when there is a lag to the effect (e.g. chronic effects). (This discounting for lag from current exposure to effect is a separate point from the discounting for exposures in future years which would be part of a full cost-benefit analysis). We considered that valuing future safety benefits should only involve discounting at the pure time preference rate for utility. The Treasury's "Green Book" suggests that a plausible long term value of the pure time preference rate for utility would be around 1% per year. Sensitivity analysis could also be done using a discount rate of 6% per year (the public sector discount rate).

Adjustment factors for air pollution acute mortality

7.32 We now summarise what might be the appropriate adjustment factors to use in the context of deaths brought forward by air pollution. As mentioned previously, there are uncertainties in defining the typical characteristics of those affected and also uncertainties in the appropriate degree of adjustment. With that proviso, we suggested the following adjustments to the baseline VPF for acute effects of £2m (see Table 7.1).

Table 7.1 **Proposed Adjustment Factors to the Baseline VPF**

	Adjustment factor	Comment
Age	70%	Those affected are mainly over 65. Some empirical evidence on WTP and age.
Impaired health state - life expectancy reduced below average for age - quality of life below average for age	At most, divide by 12 for 1 year or divide twice by 12 for 1 month Assume 53% (26% to 92%)	For respiratory deaths, those affected probably only have a life expectancy of 1 month or less up to 1 year and a quality of life of about 0.4 ($\pm$ 1 SD 0.2 to 0.7). Theory supports a reduction in WTP with reduced life-expectancy but unclear by how much. Unlikely to be more than in proportion. Assume WTP also drops in proportion to quality of life but limited evidence.
Level of risk; Income; Discount rate	No adjustment	Some debate over need for adjustment for baseline level of risk.

This implies a VPF for an acute respiratory death brought forward by air pollution of somewhere below £1.4m (£2m x 70%). It is unclear how much below, but it is unlikely to be lower than £2,600 (£1.4m x 1/12 x 1/12 x 26%) for a loss in life expectancy of 1 month and a quality of life of 0.2.

7.33 This range of values is lower than the figures used in other work aimed at valuing the benefits of reductions in air pollution. A standard VPF based on reviews of all VPFs in the literature is often used (e.g., £2 million in a British Lung Foundation report (Dobson-Mouawad et al, 1998); 2.6 million ECU (£1.75m) (European Commission, 1995)). Ostro and Chestnut (1998) adjusted their VPF from the US literature to account for 85% of those affected being elderly and WTP being 25% lower in the elderly. Nonetheless, they still had a high figure of $3.6m (£2.2m). They did not adjust for reduced life expectancy or quality of life. A similar approach has been taken in a paper for the World Bank (Maddison et al, 1997) which used a value of $3.2m (£2m). Smith (1998, in press) did not suggest a specific value but did argue, as we have, that it was not appropriate to use VPFs for accidents in the air pollution context. This argument was also set out along with some suggested adjustment factors in the NERA/CASPAR report (Rowlatt et al, 1998).

7.34 We are aware that some of the deaths brought forward by air pollution are likely to be cardiovascular deaths. Information on quality of life and life expectancy of those affected by air pollution is even more uncertain than for respiratory disease. Those affected could be younger or may not notice any symptoms before a sudden death. This would mean a greater loss of quantity or quality of life than in the case of deaths from respiratory disease and so, in terms of departures from our baseline VPF, a smaller downward adjustment. Thus, the value above could be higher if the cardiovascular deaths were taken into account.

7.35 We note that the impaired health state adjustment has a major impact on the final answer and contains significant uncertainties. On one set of assumptions, the fact that most of those affected would already be in poor health and would have died in any case within the next year is assumed markedly to reduce the WTP to reduce the risk of death brought forward by air pollution. On another, since the background risk of death in this group may be high, this may make the extra months all the more valuable. (The way in which high background risk might affect WTP is unclear). The two sets of assumptions could give substantially different figures for WTP. In addition, we have limited understanding of how baseline quality of life would be expected to affect WTP. There is also debate about the possible overlap between adjustments for age and for reduced life expectancy. This all emphasises the fact that the above figures should be regarded as speculative and very much an interim solution. Direct studies of WTP for reductions in air pollution mortality risks are needed. Further work on validation of the adjustment factors would also be valuable since the improved adjustment factors could be used to predict WTP in other areas where empirical work was not yet available.

NHS costs

7.36 It is unclear whether there would be NHS savings due to reductions in deaths brought forward by air pollution. It is plausible that there is just a change in the timing of the costs of caring for dying patients within one year. On the other hand, if people's lives are extended by 1 month to 1 year, then there could be additional costs to the NHS of £200 to £2500 for their care in the meantime. It is appropriate to estimate net NHS costs as part of an analysis of financial costs and benefits although we acknowledge there could be some controversy over whether costs in extended life should be taken into account. We show the benefits with and without their inclusion. In any case, this does not have a major impact compared with the WTP figures.

Chronic exposure and mortality

7.37 It was noted above that we have had insufficient time to consider fully the quantification of chronic effects. The same is also true of the valuation of chronic risks of mortality. One option is to use the procedure outlined above for acute mortality. We considered that for chronic mortality risks the baseline air pollution VPF should still be about £2m unless it included mortality from cancer where we suggest about £2.5m to take account of people's dread of cancer over and above other diseases. Discounting (to adjust for the fact that chronic deaths will occur after a lag) would of course be required. Choice of the other adjustment factors would be more difficult given there is even less information available than for acute mortality. However, it is plausible that those affected might be more like the population as a whole and thus, there would not be the same need for downward adjustment as there was with the acute effects.

7.38 Another option we briefly considered was valuing years of life lost. It is easier to express results of the chronic studies in terms of years of life lost. We rejected the simplistic version of this (VPF divided by years lost) and started to look into more sophisticated approaches taking into account more factors than number of years alone. (We know, for example, that WTP depends on age). However, we did not have time to explore this fully.

7.39 We recommend that further thought is given to this issue at a later date. In the meantime, we feel that our procedure for adjusting a VPF for acute mortality risks may reasonably be applied, with similar caveats, to the case of chronic mortality risks.

Benefits of reducing morbidity

7.40 The benefits of reduced morbidity are made up of reductions in:

(i) public costs e.g., NHS costs

(ii) private costs e.g., for medicines

(iii) lost output if people are prevented from working through ill-health

(iv) welfare costs (reflecting the pain and discomfort of illness).

We deal with each of these below concentrating on respiratory hospital admissions - the only morbidity outcome fully quantified in the COMEAP report. However, we also consider other outcomes which may be used in sensitivity analyses or in future cost-benefit work.

NHS costs

7.41 We looked at the costs to the NHS of respiratory and cardiovascular hospital admissions. We suggest figures of about £1,400 to £2,500 for a respiratory hospital admission and about £1,500 to £1,700 for a cardiovascular admission. Of course, there is some variation around these figures and greater understanding of the types of patients affected could influence the answer. The latter figures are based on stays in medical wards and would be higher for stays on cardiology wards. We note that there is some suggestion that air pollution could affect cerebrovascular disease (e.g., strokes) and in this case costs per hospital stay would be higher.

7.42 It is likely that there are other costs to the NHS based on effects of air pollution on other outcomes such as out-patient visits, GP visits, pharmaceutical costs etc. However, there is limited information on the degree of effect that air pollution has, how this affects healthcare and the cost consequences.

Private costs, lost output

7.43 Other costs which might be attenuated by reductions in air pollution include individuals' expenses for visits to the NHS (likely to be small); costs of avertive behaviour (unknown but could include expenditures such as moving house to avoid air pollution), and losses of work output net of consumption (less likely where those affected are elderly). We consider there should be more investigation of these other costs.

Welfare costs

Gains in quality of life

7.44 One way of reflecting the impact on a patient's welfare of an episode of illness such as a respiratory hospital admission is in terms of the duration and the shift in quality of life. Although reductions in air pollution could prevent a certain number of respiratory hospital admissions, this would not return patients with underlying disease to full health. Some preliminary data suggest that, for COPD patients before admission to hospital, the quality of life score on the EQ5D scale was about 0.5 ($\pm$ 1 SD 0.2 - 0.8). There were no data available on quality of life during the deterioration which had resulted in admission. However, the rough bounds would be from a deterioration so small that it did not register on the scale to a deterioration from 0.8 to death (0). The true range in shifts of quality of life is probably less than this. For a change of quality of life score of perhaps 0 - 0.8 and a length of stay of about 8 - 14 days (0.02 - 0.04 years), this corresponds to 0 - 0.03 quality-adjusted life years lost per hospital admission. This compares with about 0.02 - 0.7 quality-adjusted life years lost per death brought forward.

7.45 We recognise that this is a very approximate calculation inferred from evidence which was not intended to derive quality of life changes from the effect of air pollution and only gives part of the picture. We reiterate the need for the further work on quality of life mentioned above in paragraph 7.17.

Willingness to pay studies

7.46 There are a few WTP studies available for respiratory outcomes, mainly from the US. However, there are no WTP studies available for a respiratory hospital admission - the only morbidity outcome quantified in the COMEAP report. We note that such a study is underway in Europe but the results will not be available in time to use in this report.

7.47 Although it is obviously preferable to have empirical studies available, we considered whether it would be possible to give a rough estimate inferred indirectly from other studies in the interim. One option was to use the suggested rule of thumb (based on US data) of multiplying the cost of illness by two. We were concerned that this might not apply to UK healthcare cost data and also that it would vary substantially by disease. Thus, although this could be used as a rough check that answers were in the right order of magnitude, we considered it best to pursue other indirect methods.

7.48 Another proposed indirect method is to use a regression equation relating WTP to quality of well-being (QWB) scores, although this too has its uncertainties. We noted some of the difficulties in selecting an appropriate symptom complex score and the lack of a smooth distribution of QWB scores from 0 - 1 in Chapter 6. In addition, the equation is based on more minor effects and there may be inaccuracies in extending it to apply to more serious effects such as respiratory hospital admissions. The QWB scores were not based on empirical evidence and it would be preferable, in the UK, to consider EQ5D scores instead. Although there clearly is a correlation, the relationship between WTP and quality of life has not been investigated directly to any extent. This would obviously be helpful and applicable more widely than to the air pollution context alone.

7.49 To use the above statistical relationship requires a QWB score for the change in quality of life due to a respiratory hospital admission. No empirical evidence on this change was available. We lack information on quality of life using the same scoring system before, during and after hospital admission. However, using a combination of empirical evidence on COPD outpatients (QWB score 0.6 ($\pm$1SD 0.5 - 0.7)) and a theoretically predicted QWB score during a hospital admission of 0.47, we estimated the WTP to avoid a hospital admission lasting 8-14 days as £170 - £735. This is less than the NHS costs and not double (£2,800 - £5,000) as would be suggested by the rule of thumb described in 7.47.

7.50 Most inferred values for WTP to avoid a respiratory hospital admission in the literature are based on this rule of thumb adjustment to United States cost of illness data. This leads to higher values (£9380 (Maddison *et al*, 1996); 6,600 ECU (£4,500) (European Commission, 1995); $14,000 (£8,600) (Ostro and Chestnut, 1998)) than we have suggested here. This is partly due to the high healthcare costs in the US but also because of the difference in analytical approach. We have based our approach on that used in a paper submitted to the World Bank (Maddison *et al*, 1997) although the figure given in this paper ($4732 (£2,900)) is higher than our figure. We have taken into account the shift in quality of life on admission to hospital rather than the absolute quality of life score during a hospital stay and this may account for our lower figure.

7.51 We suggest that cardiovascular admissions might be valued similarly to respiratory hospital admissions. Length of stay and shift in quality of life could be smaller or larger than for respiratory hospital admissions depending on the type of effect on heart disease. This is currently insufficiently understood. There is some evidence of an effect of air pollution on strokes - avoiding this would be valued more highly since large shifts in quality of life and extended lengths of stay in hospital are possible.

7.52 There are several issues arising from this interim approach to valuing morbidity. The assumption behind the regression equation is that there will be just one WTP value for a particular quality of life state whereas it will be clear from discussions in previous chapters that the value could vary according to the disease responsible for the quality of life state or according to the context of the risk (e.g., as a result of air pollution). The regression equation suggests that the same size of shift in quality of life is valued differently according to the starting point i.e., if quality of life is already poor a further small shift is regarded with more concern than the same shift from a higher quality of life. A similar point applies to duration - a further day on top of several previous days of poorer quality of life may be valued more or less highly than after just 1 day. This adds to the need for a full understanding of the exact circumstances of those affected.

Overview of results

7.53 Our suggested estimates are summarised in Table 7.2. We emphasise again that these are very much interim estimates. The comments in the table indicate briefly how the estimates were derived and this of course has been discussed extensively earlier in the report. It is important when using the estimates to ensure that it is clear that they were inferred indirectly and could be much improved. Nonetheless, the rough order of magnitude of the estimates may be useful in coming to broad conclusions about the importance of different outcomes and the importance of the effects of air pollution compared with other effects.

Table 7.2 Summary of Estimates of Measures of Importance of Benefits from Reductions in Air Pollution (1996 Prices)

Measure of importance of benefit from reduction in pollution	Per death brought forward (acute)	Per respiratory hospital admission	Comments
Gains in quality of life and life expectancy	Quality of life of 0.2 to 0.7 for perhaps 1 month to 1 year. (0.02 - 0.7 QALYs) (respiratory)	Avoided reduction in quality of life of 0 - 0.8 for 8 - 14 days (0 - 0.03 QALYs)	For hospital admissions, *change* in quality of life due to air pollution not directly studied - inferred from some empirical evidence on COPD patients before hospital admission.
NHS costs saved	None saved, added costs in extra 1 month to 1 year of life: £200 - £2500.	£1,400 - £2,500	In-patient costs only. There could be other NHS costs.
People's willingness to pay (WTP) for a small reduction in risk aggregated so as to apply per death brought forward or for avoiding a hospital admission	Upper bound: £1.4m Low estimates: 1 year £32,000 to £110,000 1 month £2,600 to £9,200 (respiratory)	£170 - £735	No direct empirical evidence. Mortality value predicted from judgements of effect of context, age and impaired health state on WTP. Respiratory hospital admission figure from relationship between WTP and quality of life using empirical evidence and predicted scores. Other inferred figures in the literature are higher.
Other		[Private costs unknown but small]	Savings in costs of avertive behaviour unknown. Lost work output n/a to elderly or seriously ill.

7.54 Given that these estimates are derived from factors which have a variety of possible values, some consideration needs to be given to how to combine them while retaining a measure of the underlying variation. The nature and variation of the underlying inputs are outlined below:

Dose-response functions: These are given as means with 95% confidence intervals (Annex 2A). This is an appropriate measure of sampling errors, although it does not measure all aspects of uncertainty. The dose-response functions are the only parameters for which confidence intervals were available. They are not used to derive the estimates in Table 7.2 but are used in calculations of benefits (see Table 7.3).

Quality of life: The baseline quality of life data were available as means and standard deviations. The sample standard deviation gives a measure of the variation in individual values, whereas for our purposes, a measure, such as the standard error, of variation in the sample mean across samples would be more appropriate. No standard error was available and there were too few studies available to use a range across studies as a measure of population variation. We have, therefore, used the standard deviation in the absence of a better alternative. There are no data on quality of life during hospital admission - we have used a theoretically predicted score (QWB) or an estimate of a plausible shift from the baseline.

Life expectancy: As mentioned previously, there are no direct data on loss of life expectancy - the values are based on judgement of the likely range. It would be possible to take a mid-point in the range as a single value but we considered that this might imply a level of accuracy similar to single values derived from firmly based averages. The use of a range would better reflect the uncertainty.

Duration of hospital admissions: This is based on the range between the duration averaged across all ages and the average duration for the over 65s. The duration of the hospital admissions associated with air pollution is unknown but is probably somewhere between the two values. Again, we considered the range better reflected the uncertainties.

7.55 These inputs then need to be combined into overall estimates. It would be possible to present "middle" estimates but, while these might appear to be more accurate than the ranges, they actually involve an additional assumption as to what mid-point to choose in the underlying inputs. There was no case where it was possible to combine a series of well-established mean values. We took the view that the overall estimates were best expressed as ranges using the extreme values of the ranges in the underlying inputs (e.g., low quality of life with low life expectancy) and this is what is given in Table 7.3. However, for illustration, some plausible "middle" estimates would be as follows. Assuming a mid-point for loss of life expectancy of 6 months and a mean quality of life of 0.4 would give a VPF of £32,000 and a loss of 0.2 quality-adjusted life years per death brought forward. Assuming a mid-point for duration of hospital admissions of 11 days and a change in QWB score from 0.6 to 0.47 would give a WTP value of £530 per hospital admission avoided. Assuming the same duration, a mean EQ5D baseline score of 0.5 and a mid-point of 0.25 between a possible range of shifts leading to hospital admission of 0 (no change) to 0.5 (change from baseline to a state as bad as death), would give a loss of 0.0075 quality-adjusted life years per hospital admission.

7.56 As would be expected, in terms of quality-adjusted life years and WTP, respiratory hospital admissions are regarded as less important than deaths brought forward. The lower end of the range for deaths brought forward is close to that for hospital admissions. This is not surprising since the same size of shift in clinical condition which could lead to a hospital admission in someone who was moderately ill could lead to a loss of 1 months life expectancy in someone who was severely ill. The upper end of the range is significantly higher for deaths brought forward than hospital admissions.

7.57 The difference in severity between deaths brought forward and hospital admissions is not clearly translated into differences in NHS costs. We have no evidence on how death from air pollution affects NHS costs - we have assessed how it may shorten life, but this is the only way in which we have inferred an impact on the NHS. It is also unclear what proportion of the hospital admissions are additional - if most hospital admissions are merely "brought forward" then the savings in NHS costs will be overstated.

7.58 We note that air pollution is likely to have an effect on a wider range of outcomes than are covered here. The effects of chronic exposure are potentially substantial (see paragraph 7.65). Our initial thinking suggests that the importance attached to other morbidity outcomes is likely to be low compared with respiratory hospital admissions. The key question here is whether the number of people affected or the frequency of the health outcome (both plausibly greater than for hospital admissions) would outweigh these lower values. There are too many uncertainties to resolve this at present.

Overview of effect of different pollutants

7.59 We do not aggregate the benefits of a particular reduction in air pollution in this report - this will be done in the full cost-benefit analysis to which this report will contribute. However, we can express the benefits per unit concentration of pollutant. This takes into account whether there is, say, more effect on respiratory hospital admissions than deaths and also whether one pollutant is more potent than another. However, it must be emphasised that this does not indicate whether a particular pollutant has more effect overall since this will depend on the actual concentrations of pollutants to which people are exposed. It should also be emphasised that, although it is a reasonable assumption that the benefits from a reduction in a pollutant would be linear since the dose-response functions for individual pollutants are linear with no threshold, a non-linear response cannot be ruled out because of possible complex interactions within the mixture of pollutants. It should be noted that the WTP estimates do not represent actual costs that have been incurred.

7.60 The calculations below use the dose-response functions with a range of uncertainty from Table 2A.1 in Annex 2A and the baseline rates specified in paragraph 2.35. The quantified health outcomes are then combined with the estimates in Table 7.2 above. Where there are ranges in both the health outcome and the estimates in Table 7.2, the lowest of each range and the highest of each range have been combined to give the ranges below. However, it is inappropriate to add figures which relate to 1 months loss of life expectancy to figures which relate to 1 years loss of life expectancy. A worked example for calculating the health outcomes per unit concentration for PM_{10} was given in paragraph 2.35. To continue with this example, the number of deaths needs to be multiplied by the WTP estimate for deaths brought forward from Table 7.2. Both the numbers of deaths and the WTP estimates are given as ranges. The lowest number of deaths suggested is 270 and the lowest WTP estimate suggested is £2,600. Thus, the lowest overall estimate would be 270 x £2,600 = £0.7m. This is then added to the lowest estimated NHS savings (180 hospital admissions x £1,400 = £0.25m) and the lowest estimated WTP for avoiding hospital admissions (180 x £170 = £0.03m). This gives an overall total of £0.98m. The process is then repeated using the highest estimates in each case. If NHS costs incurred in extended life are included then they need to be subtracted from the total benefits (ensuring that NHS costs in an added *year* of life are not subtracted from WTP for a loss in life expectancy of 1 *month*). For example, the overall total of £0.98m (based on 1 month) net of NHS costs (270 x £200 = £0.05m) for 1 extra month would be £0.93m. The results of all the calculations are given in Table 7.3. For quality-adjusted life years, a similar process was followed and the results are given in Table 7.4.

Table 7.3 **Summary of Estimates of Measures of Importance of Benefits per Unit Concentration of Pollutant per year or per summer (1996 prices)**

	Health outcome[a] and measure of importance[b] per $\mu g/m^3$ reduction of pollutant		
	PM_{10} in GB urban population per year	Sulphur dioxide in GB urban population per year	Ozone in GB urban and rural population summer only
Reduction in deaths brought forward (all cause)(acute)	340 (270 - 385)	270 (225-315)	170 (70-225)
WTP for reduction[c]	£0.7m - £540m	£0.58m - £440m	£0.18m - £315m
Reduction in respiratory hospital admissions	280 (180-390)	180 (-53 to 320)	145 (50 - 245)
NHS savings	£0.25m - £0.98m	-£0.07m - £0.8m	£0.13m - £0.61m
WTP for reduction	£0.03m - £0.29m	-£0.01m to £0.24m	£0.01m - £0.18m
Total NHS savings	£0.25m - £0.98m	-£0.07m - £0.8m	£0.13m - £0.61m
Total WTP	£0.73m - £540m	£0.57m - £440m	£0.19m - £315m
Total benefits	**£0.98m - £540m**	**£0.5m - £440m**	**£0.32m - £315m**
NHS costs incurred[d]	£0.05m - £0.96m	£0.05m - £0.79m	£0.01m - £0.56m
Total benefits net of NHS costs	**£0.93m - £540m**	**£0.45m to £440m**	**£0.31m - £315m**

Notes to Table 7.3

[a] Calculated from the mean and standard deviation of the dose-response functions (percentage increase in health outcomes per unit concentration of pollutant) in the COMEAP report, baseline rates for the health outcomes and numbers in the relevant population (see Chapter 2 and Annex 2A). Figures would be higher if the Northern Ireland population were included and lower for hospital admissions associated with ozone if there is a threshold of 50 ppb.

[b] Given as range from highest estimate (top end of range for health outcome and top end of relevant range in Table 7.2) to the lowest estimate (bottom end of the relevant ranges). The figure of -53 for hospital admissions and sulphur dioxide represents statistical variability - a beneficial effect of sulphur dioxide is not supported by other evidence.

[c] Upper figure based on upper bound VPF of £1.4 m; probably an overestimate. Lower figure based on proportional adjustment according to a loss of life-expectancy of 1 month and a quality of life of 0.2.

[d] NHS costs incurred if lives exztended by 1 month to 1 year.

Table 7.4 **Summary of Estimates of Gains in Quality-Adjusted Life Years per Unit Concentration of Pollutant per Year or per Summer**

	Health outcome[a] and measure of importance[b] per $\mu g/m^3$ reduction of pollutant		
	PM$_{10}$ in GB urban population per year	Sulphur dioxide in GB urban population per year	Ozone in GB urban and rural population summer only
Reduction in deaths brought forward (all cause)(acute)	340 (270 - 385)	270 (225-315)	170 (70-225)
QALYs gained from reductions in deaths brought forward (all cause)	5 - 270	5 - 220	1 -160
QALYs gained from reductions in hospital admissions	0 - 12	0 - 9.6	0 - 7.4
Total QALY gains	5 - 282	5 - 230	1 - 167

[a] Calculated from the mean and standard deviation of the dose-response functions (percentage increase in health outcomes per unit concentration of pollutant) in the COMEAP report, baseline rates for the health outcomes and numbers in the relevant population (see Chapter 2 and Annex 2A). Figures would be higher if the Northern Ireland population were included and lower for hospital admissions associated with ozone if there is a threshold of 50 ppb.

[b] Given as range from highest estimate (top end of range for health outcome and top end of relevant range in Table 7.2) to the lowest estimate (bottom end of the relevant ranges). The figure of -53 for hospital admissions and sulphur dioxide represents statistical variability - a beneficial effect of sulphur dioxide is not supported by other evidence.

7.61 Table 7.3 demonstrates that benefits from reducing deaths brought forward continue to dominate the results after taking the dose-response functions into account although this is less marked at the bottom of the range. The total benefits range over 3 orders of magnitude, reflecting the uncertainties involved. The results suggest that the benefits from reductions in particles are greater than the benefits from reductions in sulphur dioxide which are in turn greater than those for ozone. However, the ranges do overlap and the ozone calculations are done on a slightly different basis. Table 7.4 shows a similar ranking of pollutants in terms of quality-adjusted life years.

Uncertainty

7.62 We have mentioned areas of uncertainty throughout the report and highlight some of the major ones here. Uncertainties in the epidemiological studies are also discussed in the COMEAP quantification report. It is clear that the overall benefits are dominated by the estimates for deaths brought forward so consideration of the uncertainties in these estimates is most important. The basic cause of the uncertainty is the absence of empirical valuation studies in this area. We have suggested adjustment factors to be applied to values obtained from other areas. A fundamental issue is whether to adjust the values at all - there are possible ethical arguments that the weak should be protected as much as the strong for instance (but also counter-arguments that effort should be concentrated where the greatest benefit can be achieved and where people's preferences lie). This issue is not unique to air pollution. Although there could be debate over adjustment for age, there is at least empirical evidence that people's WTP does decrease with age, beyond middle age. The evidence is weaker for adjustments for life expectancy and weakest for the adjustment for quality of life. In addition, as discussed in Chapter 4, some theoretical models (but not others) suggest that the high baseline level of risk in those with low life expectancy could increase WTP and may attenuate the downward adjustment for impaired health state. It is for these reasons that we have given a wide range for the WTP estimates starting from an upper bound of £1.4m (with no adjustment for reduced life-expectancy or quality of life). We have also given some possible lower estimates - although this is probably a lower bound for life expectancy adjustment, it is unclear whether it is a lower bound for the quality of life adjustment. Unfortunately, we have insufficient information to give a single "best estimate".

7.63 The choice of appropriate adjustment factors for the air pollution context relies on evidence on the characteristics of those affected being available. At present the evidence is limited. The WTP value is based on an average loss of life expectancy of from 1 month to 1 year but as mentioned previously the average loss of life expectancy is uncertain. (Average losses in life expectancy less than 1 month are possible but it is unlikely that the *average* will be as low as one or two days and it also unclear that people would regard losses of life expectancy measured in days in the same way as years). The choice of the appropriate assumption for loss of life expectancy has a large impact on WTP. There are also uncertainties in quality of life estimates but this would have a smaller effect.

7.64 The proposed adjustment factor for impaired health state is based on respiratory deaths but the numbers of deaths brought forward actually include cardiovascular deaths as well (probably in roughly equal proportions). The effect on cardiovascular disease potentially involves greater losses in quality of life and life expectancy. This would be expected to increase the WTP values to some extent.

7.65 A major uncertainty concerns the chronic effects. There is evidence for a chronic effect on mortality from the United States although there is uncertainty over whether this evidence is transferable to the UK and over the size of the public health impact. This could be substantial in terms of numbers of deaths, for particles at least, and the WTP values could be higher. The age of those affected is unclear and, although most of the deaths are cardiorespiratory deaths, there is some indication of an effect on lung cancer. Other pollutants not covered here also have effects on cancer which would probably be valued more highly but there are great difficulties in quantifying effects due to uncertainties in extrapolation from high exposures in occupational studies (the only evidence available) to exposures at environmental levels.

7.66 Variation in morbidity effects is likely to be less marked and has a minimal effect on total benefits compared with mortality effects. If nitrogen dioxide and, possibly carbon monoxide, were more clearly shown to have separate effects from other pollutants then this would add to the morbidity effects. However, unless large concentration changes were involved this would not be expected to cause a major increase. The possible inclusion of cardiovascular admissions for PM_{10} would probably give results of the same order as for respiratory admissions. Confirmation of an effect on strokes could have a larger impact due to the longer duration of hospital admissions and large effects on quality of life. The inclusion of more minor effects would only have a large impact if greater numbers of people were affected compared with the numbers of people hospitalised.

7.67 Although some of the individual variations discussed may be quite small their cumulative impact may be more significant. Therefore, the results should be judged as broad indications only.

Conclusions and further work

7.68 In summary, we have:

(i) concluded that monetary valuation is the approach that can be used to determine most clearly whether the benefits exceed the costs (or vice versa) in a fully quantified cost-benefit analysis but that it can usefully be complemented with investigations of quality of life and life expectancy, multi-criteria approaches and data on people's more qualitative views of the relative importance of different benefits;

(ii) given an indication of the gains in quality of life and life expectancy which might be achieved by reductions in pollution (see Tables 7.2 and 7.4) based on the limited evidence which was available;

(iii) noted the lack of direct empirical evidence on monetary valuation of the reduction in risk of deaths brought forward by air pollution and of respiratory hospital admissions;

(iv) concluded that it is not appropriate to apply empirical evidence on monetary valuation of the reduction in risk of deaths in accidents directly and without adjustment to the air pollution context;

(v) suggested what WTP values might be expected for deaths brought forward by air pollution (see Tables 7.2 and 7.3) based on knowledge of the effect on WTP of various factors such as the age and impaired health state of those affected. Our suggested estimates for those who are of advanced years and in an already impaired health state cover a wide range and include values which are substantially lower than those found for deaths in accidents;

(vi) suggested what WTP values might be expected for reductions in hospital admissions (see Tables 7.2 and 7.3) based on a statistical relationship between WTP and quality of life from other studies;

(vii) estimated healthcare costs for respiratory hospital admissions (see Tables 7.2 and 7.3);

(viii) suggested some sensitivity analyses for outcomes other than those quantified in the COMEAP report and discussed possible valuation estimates and healthcare costs for some of these endpoints should stronger evidence be available in future;

(ix) given initial consideration to chronic exposure and mortality and noted that the impact of chronic effects is potentially large and valuation estimates potentially high. Were this to be confirmed, it might dominate any future valuation exercise.

Further work

7.69 It will be clear from the above that the empirical evidence available to us was limited. Our suggested estimates are, therefore, very speculative although we hope that they give a broad indication of the possible values. Further work to provide more reliable estimates is important and is outlined below.

Epidemiology

7.70 We endorse the further epidemiological work recommended in the COMEAP report on the period of life lost, the impact of long term exposure, research on groups at special risk and effects on outcomes other than deaths and hospital admissions. This is important not just for identifying the health problems but also because better understanding of the characteristics of those affected will improve the WTP estimates. We note, in particular, the need for better understanding of the effect on heart disease, the need for studies of minor outcomes to define symptoms and subjects to match the population statistics available and the need to better define the impact of the chronic effects.

Quality of life

7.71 We suggest more work is done on the quality of life of patients with respiratory and cardiovascular disease and the effects of air pollution on this using both generic and disease-specific measures.

People's attitudes to air pollution risks

7.72 More research is needed on (i) the levels of knowledge and understanding people currently hold on the causes and effects of air pollution, (ii) what extra knowledge people might need to make informed decisions about their personal health protection with respect to air pollution and (iii) what processes of deliberation and information provision might help people to construct stable and consistent preferences regarding air pollution risk reduction.

NHS and other costs

7.73 Better understanding of the epidemiological evidence will help in refining the estimates of NHS costs. Further investigation of private costs (travel costs, avertive behaviour etc) and any impact on lost output is also needed.

Monetary valuation

7.74 We recommend empirical studies of people's WTP to reduce the risk of their death being brought forward by air pollution. As noted in Chapter 3, successful application of the contingent valuation approach requires considerable care and circumspection and experience (Chilton *et al*, 1998) suggests that a judicious blend of the contingent valuation and relative valuation approaches may offer the most promising way forward.

7.75 Similarly, we recommend empirical work on WTP for reductions in respiratory hospital admissions and other morbidity effects which may be affected by air pollution.

7.76 In addition, for valuation of mortality, we consider that further work on the validation of the adjustment factors we have suggested would be valuable. Although prediction of the WTP value will be unnecessary if empirical evidence is available, this will help in understanding the reasons behind people's WTP and will be helpful in other areas.

7.77 Further thought on the valuation of chronic effects is needed. Empirical valuation work would be useful but may need to await better epidemiological evidence.

7.78 Although not specific to the air pollution context, it is clear that better understanding of the relationship between WTP and quality of life is important. This will help interpret any differences between the WTP approach and the quality-adjusted life year approach which are both used to assess the importance of benefits of government policies. Willingness to pay for reductions in both mortality and morbidity risks is likely to be influenced by the changes in quality of life and life expectancy involved, but it may also be affected by other considerations.

Final word

7.79 This report assesses the relative importance of the various health benefits of reductions in air pollution. It is a first step and there are many uncertainties but we have identified further work to improve the situation. We have noted the steps in our approach where ethical choices are involved and others' opinions may differ. Nonetheless, we believe that this report, and the cost-benefit analysis to which it will contribute, will help illuminate the positive and negative implications of air quality policy and the balance between them.

References

Chilton, S., Covey, J., Hopkins, L., Jones-Lee, M., Loomes, G., Pidgeon, N. and Spencer, A. (1998) *New research results on the valuation of preventing fatal road accident casualties.* In: Road Accidents Great Britain 1997. London: The Stationery Office.

Dobson-Mouawad, D., Dobson-Mouawad, L. and Pearce, D. (1998) *Transport and Pollution - the Health Costs*, British Lung Foundation. London: British Lung Foundation.

European Commission. (1995) *Externe E - Externalities of Energy. Volume 2. Methodology.* Luxembourg: Office for Official Publications of the European Communities.

Maddison, D., Pearce, D., Johansson, O., Calthrop, E., Litman, T. and Verhoef, E. (1996) *Blueprint 5. The True Costs of Road Transport.* London: Earthscan Publications Ltd.

Maddison, D., Hughes, G., Lvovsky, K. and Pearce, D. (1997) *Measuring the effect of air pollution in large urban conurbations.* Paper submitted to the World Bank.

Ostro, B. and Chestnut, L. (1998) Assessing the health benefits of reducing particulate matter air pollution in the United States. Environ. Res. **76**: 94-106.

Rowlatt, P., Jones-Lee, M., Spackman, M., Loomes, G. and Jones, S. (1998) *Valuation of Deaths from Air Pollution.* (Report prepared for the Department of Environment, Transport and the Regions and the Department of Trade and Industry). London: National Economic Research Associated.

Smith, A.E. (1998) *Valuing Mortality Risks Associated with Air Pollution.* Washington, DC: DFI Aeronomics.

APPENDIX 1.

Glossary of Terms and Abbreviations

Acute myocardial infarction	The myocardium is the muscular tissue of the heart. If the blood supply to the myocardium is blocked areas of the muscle undergo infarction, ie, damage due to a lack of oxygen. An acute myocardial infarction is a heart attack
ALARP	As low as reasonably practicable
APHEA	Short Term Effects of Air Pollution on Health: a European Approach (using epidemiological time series data) was initiated and funded in the framework of the EC Environment 91-94 Programme. Its main objective is to provide quantitative estimates, using standardised methods, of the short term effects of air pollution in Europe, with data from 15 large cities representing various social, cultural, environmental and air pollution situations
Black Smoke (BS)	Non-reflective (dark) particulate matter, measured by the smoke stain method
Bronchodilator	An agent that increases the diameter of the air passages (bronchi) of the lungs
CASPAR	Centre for the Analysis of Safety Policy and Attitudes to Risk (University of Newcastle upon Tyne)
Cardinal	A mathematical term; on a cardinal scale if one number is double another then it follows that there is twice as much of whatever is being measured - e.g., 2 μg is twice as heavy as 1 μg
Cardiovascular	Relating to the heart and circulatory system. In the context of cardiovascular hospital admissions, used more restrictively to relate just to the heart
Cerebrovascular	Relating to the blood vessels of the brain
Chamber studies	Studies involving exposure of volunteers to controlled atmospheres in experimental chambers
Coefficient of haze	A measure of the optical absorption density of airborne particles, closely related to Black Smoke
All-cause mortality	Deaths from any cause
CIPFA	Chartered Institute for Public Finance and Accountancy
CO	Carbon monoxide
Cohort studies	Studies in which a group or "cohort" of people are followed over time to see whether they develop a disease in response to exposure to the factor of interest

COMEAP		Committee on the Medical Effects of Air Pollutants
Confidence intervals or limits		A range within which the true value of a population characteristic might be expected to lie.
Confounding factors		In observational studies of populations it is not possible to avoid variation in factors other than the exposure under investigation. Some of these other factors may also influence the health effect being studied and confound the association between exposure and effect. For example, deaths are higher in cold weather. If pollution is also higher in cold weather, then pollution could appear to increase deaths even if it did not in fact do so.
Congestive heart failure		Inadequate pumping action of the heart leading to back pressure of blood in the lungs and liver
Conjoint Analysis		A form of multi-criteria analysis in which people are asked to rate different scenarios defined according to a variety of criteria. It aims to find which criteria matter most.
Contingent Valuation/ Stated Preference		Willingness to pay for something which is non-marketed as derived from people's responses to questions about preferences for various combinations of situations and/or controlled discussion groups.
COPD		Chronic obstructive pulmonary disease (often taken to mean chronic bronchitis and emphysema)
Cost-Benefit Analysis (CBA)		The most comprehensive form of economic appraisal which seeks to make costs and benefits commensurate by quantifying them as far as possible in monetary terms (including items for which the market does not provide a satisfactory measure of economic value)
Cost-Effectiveness Analysis (CEA)		The comparison of alternative ways of producing the same or similar outputs, which are not necessarily given a monetary value
DETR		Department of the Environment, Transport and the Regions
DH		Department of Health
Discount rate		The annual percentage rate at which the present value of a future pound, or other unit of account, is assumed to fall away through time
Dose-response functions		A mathematical relationship between the amount of exposure to a substance (the dose) and the response to it. For example, the increase in hospital admissions for a particular change in pollutant concentration
EC		European Commission or European Community
Epidemiology		The study of how often diseases occur in different groups of people and why.
EuroQol/EQ5D		A scale for measuring health-related quality of life (which has 5 dimensions)
ex post		The result; after the event

ex ante	Expected or intended; before the event
ExternE	Externalities of Energy - a European project examining the external costs of the use of a wide range of different fuels
GB	Great Britain
GDP - gross domestic product	The total output produced in an economy (annually)
GPs	General practitioners
Gross Output/Human Capital Approach	A method for estimating the value of a human life in terms of the value of future output lost if the person were to die
ICD codes	International Classification of Diseases codes
Income Elasticity	The proportionate change in effect produced by a proportionate change in income
Ischaemic heart disease	Heart disease caused by inadequate blood flow (due to constriction or blockage of blood vessels) to part of the heart
Leukaemia	Cancer of the bone-marrow leading to increased numbers of abnormal white blood cells
Lifetables	Tables showing life expectancy at different ages
Lost Output	Reduction in output produced - e.g., if someone were unable to work
Lymphoma	Cancer arising from lymphoid tissue (lymph nodes, spleen, thymus and sometimes bone marrow)
Marginal Utility of Wealth	The extra utility from a small change in wealth
MCDA	Multi-Criteria Decision Analysis. A form of multi-criteria analysis designed to help structure complex decisions according to the probable consequences of alternative scenarios
MCLs	Maximum concentration limits
Multicriteria Analysis	A form of analysis where factors are analysed on non-commensurate scales and the values of the decision-maker or some other group are used
NERA	National Economic Research Associates
NHS	National Health Service
NO_2	Nitrogen dioxide
O_3	Ozone
Ordinal	A mathematical term; on an ordinal scale the factors are simply ranked in order - e.g., first, second, third, etc
PM_{10}	Particulate matter less than 10 μm aerodynamic diameter (or, more strictly, particles which pass through a size selective inlet with a 50% efficiency cut-off at 10 μm aerodynamic diameter)

ppb	Parts per billion
QALY	A measure of health status in terms of the quality of life associated with a state of health, and the number of years for which that health status is enjoyed
QWB	"Quality of Well Being": a scale for measuring health-related quality of life
Relative Valuation	Willingness to pay for something which is non-marketed as derived from people's responses to questions about relative preferences for various combinations of situations and/or controlled discussion groups
Respiratory effect	Effect on the lungs
Revealed Preference	Willingness to pay for something which itself is non-marketed, as revealed by other expenditure choices
Sensitivity analysis	Analysis of the effect of varying important variables on an overall answer. It can show whether the final answer is particularly sensitive to inclusion, exclusion or changes in specific variables
SGRQ	"St George's Respiratory Questionnaire": a scale for measuring health-related quality of life of respiratory patients
SO_2	Sulphur dioxide
Standard Deviation (SD)	A measure of the variability of a set of observations around the sample mean
Standard Error	A measure of the uncertainty of, for instance, the sample mean. (Repeat samples may not give exactly the same mean). Used to derive a confidence interval
Time series studies	Studies of health of defined populations over a specified period of time
$\mu g/m^3$	A millionth of a gram in a cubic metre of air
UK	United Kingdom
Utility/Disutility	The value which people attach to goods or services
US EPA	United States Environmental Protection Agency
VoSL	Collective willingness to pay to reduce mortality risk by those at risk (aggregated up to the level where one fewer death would be expected)
VPF	Value of prevention of a statistical fatality (an alternative term for the VOSL)
WHO	World Health Organisation
WTP/WTA	Willingness to pay/willingness to accept. Economic concepts which refer to the value that people place on goods or services by reference to their preferences for receiving goods or services, or for accepting compensation if the goods or services are lost

APPENDIX 2
Ad-Hoc Group on Economic Appraisal of Health Effects of Air Pollution

Membership

Chairman	Mr N J Hartley, MA, BPhil	
Members	Professor J G Ayres, BSc, MD, FRCP, FRSA	
	Professor M J Buxton, BA (Soc Sci)	
	Dr A Jones, BA, MSc, DPhil	
	Professor M Jones-Lee, BEng, DPhil	
	Dr D Maddison, MA, MSc, PhD	
	Professor A Markandya, BA, MSc, PhD, FRSA	
	Dr N Pidgeon, BA, PhD	
	Ms M Postle, BSc, MSc	
	Dr P Rowlatt, BA, PhD, MSc	
Secretariat	Dr H Walton, BSc, DPhil	
	Miss J Cumberlidge, BSc, MSc	
	Mr J Henderson, BA, MSc	

Acknowledgement of those giving advice

Professor P G J Burney, MA, MD, MRCP, FFPHM
Mr F Hurley, MA

Observers		
	Ms C Cottingham	(DETR)
	Mr C Gadsden	(Treasury)
	Mr M Keoghan	(DTI)
	Mr L Patten	(DETR)
	Ms S Roper	(DTI)
	Ms A Rowlatt	(Treasury)
	Dr J Steedman	(DETR)

APPENDIX 3

Ad-Hoc Group on Economic Appraisal of Health Effects of Air Pollution

Members' Interests

	Personal Interest		Non-Personal Interest	
Member	Company	Interest	Company	Interest
Mr N J Hartley (Chairman)	British Airways plc	Shareholder	Oxford Economic Research Associates Ltd	Provision of economic advice
	Rolls Royce Ltd	Shareholder		"
				"
			BG plc	"
			PowerGen plc	"
			United Utilities plc	"
			BNFL	"
			ScottishPower plc	"
			English Welsh and Scottish Railways	"
Prof J G Ayres	3M	Medical Adviser	Hoechst	Research Grant
			Byk Gulden	Research Grant
	Schering Plough	Medical Adviser	BLF/Fisons	Research Grant
			Allen & Hansburys	Research Grant
	Zeneca	Medical Adviser	Smith Kline Beecham	Research Grant
	MSD	Medical Adviser	Glaxo	Research Grant
			Astra	Research Grant
			Merck Sharp & Dohme	Research Grant
			Zeneca	Research Grant
			DH	Research Grant
			DETR	Research Grant
Prof P Burney	Bibby Line	Shareholder	NAC/Glaxo	Research Grant
Prof M J Buxton	Astra Draco	Consultant	DH	Research Grants
	Searle Division of Monsanto plc	Consultant		
	Lilly Industries Ltd	Consultant		
	Rhône-Poulenc Rorer	Consultant		
	Schering AG	Consultant		
Dr A Jones	Abbey National	Shareholder	Zeneca	Research Grant
			Pfizer	Research Grant

	Personal Interest		Non-Personal Interest	
Member	Company	Interest	Company	Interest
Prof M Jones-Lee	Railtrack British Petroleum Mass Transit Railway Corporation, Hong Kong University of Hong Kong EHS Consultants, Ltd, Hong Kong New Zealand Land Transport Safety Authority Halifax plc	Consultant Consultant Consultant Consultant Consultant Consultant Shareholder	HSE DETR Home Office HM Treasury	Research Grant Research Grant Research Grant Research Grant
Dr D Maddison	NONE	NONE	NONE	NONE
Prof A Markyanda	Metroeconomica Ltd Bath University National Power	Director Employee Consultant	Department of Health EC: DGXII	Research Grant (with IOM) Research Grant
Dr N Pidgeon	NONE	NONE	NONE	NONE
Ms M Postle	NONE	NONE	International Council on Metals and the Environment	Related to receipt of research contracts
Dr P Rowlatt	Nigen Premier Power Coolkeeragh MSD Marsh McLennan Brierley Lloyds TSB St Ives BICC Stakis Cable & Wireless	Consultant Consultant Consultant Consultant Shareholder Shareholder Shareholder Shareholder Shareholder Shareholder Shareholder	NONE	NONE
J F Hurley	Institute of Occupational Medicine	Employee	Institute of Occupational Medicine	Various research grants related to air pollution